D1713472

A UNIFIED SIGNAL ALGEBRA APPROACH TO TWO-DIMENSIONAL PARALLEL DIGITAL SIGNAL PROCESSING

PURE AND APPLIED MATHEMATICS

A Program of Monographs, Textbooks, and Lecture Notes

MONOGRAPHS AND TEXTBOOKS IN PURE AND APPLIED MATHEMATICS

1. *K. Yano,* Integral Formulas in Riemannian Geometry (1970)
2. *S. Kobayashi,* Hyperbolic Manifolds and Holomorphic Mappings (1970)
3. *V. S. Vladimirov,* Equations of Mathematical Physics (A. Jeffrey, ed.; A. Littlewood, trans.) (1970)
4. *B. N. Pshenichnyi,* Necessary Conditions for an Extremum (L. Neustadt, translation ed.; K. Makowski, trans.) (1971)
5. *L. Narici et al.,* Functional Analysis and Valuation Theory (1971)
6. *S. S. Passman,* Infinite Group Rings (1971)
7. *L. Dornhoff,* Group Representation Theory. Part A: Ordinary Representation Theory. Part B: Modular Representation Theory (1971, 1972)
8. *W. Boothby and G. L. Weiss, eds.,* Symmetric Spaces (1972)
9. *Y. Matsushima,* Differentiable Manifolds (E. T. Kobayashi, trans.) (1972)
10. *L. E. Ward, Jr.,* Topology (1972)
11. *A. Babakhanian,* Cohomological Methods in Group Theory (1972)
12. *R. Gilmer,* Multiplicative Ideal Theory (1972)
13. *J. Yeh,* Stochastic Processes and the Wiener Integral (1973)
14. *J. Barros-Neto,* Introduction to the Theory of Distributions (1973)
15. *R. Larsen,* Functional Analysis (1973)
16. *K. Yano and S. Ishihara,* Tangent and Cotangent Bundles (1973)
17. *C. Procesi,* Rings with Polynomial Identities (1973)
18. *R. Hermann,* Geometry, Physics, and Systems (1973)
19. *N. R. Wallach,* Harmonic Analysis on Homogeneous Spaces (1973)
20. *J. Dieudonné,* Introduction to the Theory of Formal Groups (1973)
21. *I. Vaisman,* Cohomology and Differential Forms (1973)
22. *B.-Y. Chen,* Geometry of Submanifolds (1973)
23. *M. Marcus,* Finite Dimensional Multilinear Algebra (in two parts) (1973, 1975)
24. *R. Larsen,* Banach Algebras (1973)
25. *R. O. Kujala and A. L. Vitter, eds.,* Value Distribution Theory: Part A; Part B: Deficit and Bezout Estimates by Wilhelm Stoll (1973)
26. *K. B. Stolarsky,* Algebraic Numbers and Diophantine Approximation (1974)
27. *A. R. Magid,* The Separable Galois Theory of Commutative Rings (1974)
28. *B. R. McDonald,* Finite Rings with Identity (1974)
29. *J. Satake,* Linear Algebra (S. Koh et al., trans.) (1975)
30. *J. S. Golan,* Localization of Noncommutative Rings (1975)
31. *G. Klambauer,* Mathematical Analysis (1975)
32. *M. K. Agoston,* Algebraic Topology (1976)
33. *K. R. Goodearl,* Ring Theory (1976)
34. *L. E. Mansfield,* Linear Algebra with Geometric Applications (1976)
35. *N. J. Pullman,* Matrix Theory and Its Applications (1976)
36. *B. R. McDonald,* Geometric Algebra Over Local Rings (1976)
37. *C. W. Groetsch,* Generalized Inverses of Linear Operators (1977)
38. *J. E. Kuczkowski and J. L. Gersting,* Abstract Algebra (1977)
39. *C. O. Christenson and W. L. Voxman,* Aspects of Topology (1977)
40. *M. Nagata,* Field Theory (1977)
41. *R. L. Long,* Algebraic Number Theory (1977)
42. *W. F. Pfeffer,* Integrals and Measures (1977)
43. *R. L. Wheeden and A. Zygmund,* Measure and Integral (1977)
44. *J. H. Curtiss,* Introduction to Functions of a Complex Variable (1978)
45. *K. Hrbacek and T. Jech,* Introduction to Set Theory (1978)
46. *W. S. Massey,* Homology and Cohomology Theory (1978)
47. *M. Marcus,* Introduction to Modern Algebra (1978)
48. *E. C. Young,* Vector and Tensor Analysis (1978)
49. *S. B. Nadler, Jr.,* Hyperspaces of Sets (1978)
50. *S. K. Segal,* Topics in Group Kings (1978)
51. *A. C. M. van Rooij,* Non-Archimedean Functional Analysis (1978)
52. *L. Corwin and R. Szczarba,* Calculus in Vector Spaces (1979)
53. *C. Sadosky,* Interpolation of Operators and Singular Integrals (1979)

54. *J. Cronin,* Differential Equations (1980)
55. *C. W. Groetsch,* Elements of Applicable Functional Analysis (1980)
56. *I. Vaisman,* Foundations of Three-Dimensional Euclidean Geometry (1980)
57. *H. I. Freedan,* Deterministic Mathematical Models in Population Ecology (1980)
58. *S. B. Chae,* Lebesgue Integration (1980)
59. *C. S. Rees et al.,* Theory and Applications of Fourier Analysis (1981)
60. *L. Nachbin,* Introduction to Functional Analysis (R. M. Aron, trans.) (1981)
61. *G. Orzech and M. Orzech,* Plane Algebraic Curves (1981)
62. *R. Johnsonbaugh and W. E. Pfaffenberger,* Foundations of Mathematical Analysis (1981)
63. *W. L. Voxman and R. H. Goetschel,* Advanced Calculus (1981)
64. *L. J. Corwin and R. H. Szczarba,* Multivariable Calculus (1982)
65. *V. I. Istrăţescu,* Introduction to Linear Operator Theory (1981)
66. *R. D. Järvinen,* Finite and Infinite Dimensional Linear Spaces (1981)
67. *J. K. Beem and P. E. Ehrlich,* Global Lorentzian Geometry (1981)
68. *D. L. Armacost,* The Structure of Locally Compact Abelian Groups (1981)
69. *J. W. Brewer and M. K. Smith, eds.,* Emmy Noether: A Tribute (1981)
70. *K. H. Kim,* Boolean Matrix Theory and Applications (1982)
71. *T. W. Wieting,* The Mathematical Theory of Chromatic Plane Ornaments (1982)
72. *D. B.Gauld,* Differential Topology (1982)
73. *R. L. Faber,* Foundations of Euclidean and Non-Euclidean Geometry (1983)
74. *M. Carmeli,* Statistical Theory and Random Matrices (1983)
75. *J. H. Carruth et al.,* The Theory of Topological Semigroups (1983)
76. *R. L. Faber,* Differential Geometry and Relativity Theory (1983)
77. *S. Barnett,* Polynomials and Linear Control Systems (1983)
78. *G. Karpilovsky,* Commutative Group Algebras (1983)
79. *F. Van Oystaeyen and A. Verschoren,* Relative Invariants of Rings (1983)
80. *I. Vaisman,* A First Course in Differential Geometry (1984)
81. *G. W. Swan,* Applications of Optimal Control Theory in Biomedicine (1984)
82. *T. Petrie and J. D. Randall,* Transformation Groups on Manifolds (1984)
83. *K. Goebel and S. Reich,* Uniform Convexity, Hyperbolic Geometry, and Nonexpansive Mappings (1984)
84. *T. Albu and C. Năstăsescu,* Relative Finiteness in Module Theory (1984)
85. *K. Hrbacek and T. Jech,* Introduction to Set Theory: Second Edition (1984)
86. *F. Van Oystaeyen and A. Verschoren,* Relative Invariants of Rings (1984)
87. *B. R. McDonald,* Linear Algebra Over Commutative Rings (1984)
88. *M. Namba,* Geometry of Projective Algebraic Curves (1984)
89. *G. F. Webb,* Theory of Nonlinear Age-Dependent Population Dynamics (1985)
90. *M. R. Bremner et al.,* Tables of Dominant Weight Multiplicities for Representations of Simple Lie Algebras (1985)
91. *A. E. Fekete,* Real Linear Algebra (1985)
92. *S. B. Chae,* Holomorphy and Calculus in Normed Spaces (1985)
93. *A. J. Jerri,* Introduction to Integral Equations with Applications (1985)
94. *G. Karpilovsky,* Projective Representations of Finite Groups (1985)
95. *L. Narici and E. Beckenstein,* Topological Vector Spaces (1985)
96. *J. Weeks,* The Shape of Space (1985)
97. *P. R. Gribik and K. O. Kortanek,* Extremal Methods of Operations Research (1985)
98. *J.-A. Chao and W. A. Woyczynski, eds.,* Probability Theory and Harmonic Analysis (1986)
99. *G. D. Crown et al.,* Abstract Algebra (1986)
100. *J. H. Carruth et al.,* The Theory of Topological Semigroups, Volume 2 (1986)
101. *R. S. Doran and V. A. Belfi,* Characterizations of C*-Algebras (1986)
102. *M. W. Jeter,* Mathematical Programming (1986)
103. *M. Altman,* A Unified Theory of Nonlinear Operator and Evolution Equations with Applications (1986)
104. *A. Verschoren,* Relative Invariants of Sheaves (1987)
105. *R. A. Usmani,* Applied Linear Algebra (1987)
106. *P. Blass and J. Lang,* Zariski Surfaces and Differential Equations in Characteristic $p > 0$ (1987)
107. *J. A. Reneke et al.,* Structured Hereditary Systems (1987)
108. *H. Busemann and B. B. Phadke,* Spaces with Distinguished Geodesics (1987)
109. *R. Harte,* Invertibility and Singularity for Bounded Linear Operators (1988)

110. *G. S. Ladde et al.*, Oscillation Theory of Differential Equations with Deviating Arguments (1987)
111. *L. Dudkin et al.*, Iterative Aggregation Theory (1987)
112. *T. Okubo*, Differential Geometry (1987)
113. *D. L. Stancl and M. L. Stancl*, Real Analysis with Point-Set Topology (1987)
114. *T. C. Gard*, Introduction to Stochastic Differential Equations (1988)
115. *S. S. Abhyankar*, Enumerative Combinatorics of Young Tableaux (1988)
116. *H. Strade and R. Farnsteiner*, Modular Lie Algebras and Their Representations (1988)
117. *J. A. Huckaba*, Commutative Rings with Zero Divisors (1988)
118. *W. D. Wallis*, Combinatorial Designs (1988)
119. *W. Więsław*, Topological Fields (1988)
120. *G. Karpilovsky*, Field Theory (1988)
121. *S. Caenepeel and F. Van Oystaeyen*, Brauer Groups and the Cohomology of Graded Rings (1989)
122. *W. Kozlowski*, Modular Function Spaces (1988)
123. *E. Lowen-Colebunders*, Function Classes of Cauchy Continuous Maps (1989)
124. *M. Pavel*, Fundamentals of Pattern Recognition (1989)
125. *V. Lakshmikantham et al.*, Stability Analysis of Nonlinear Systems (1989)
126. *R. Sivaramakrishnan*, The Classical Theory of Arithmetic Functions (1989)
127. *N. A. Watson*, Parabolic Equations on an Infinite Strip (1989)
128. *K. J. Hastings*, Introduction to the Mathematics of Operations Research (1989)
129. *B. Fine*, Algebraic Theory of the Bianchi Groups (1989)
130. *D. N. Dikranjan et al.*, Topological Groups (1989)
131. *J. C. Morgan II*, Point Set Theory (1990)
132. *P. Biler and A. Witkowski*, Problems in Mathematical Analysis (1990)
133. *H. J. Sussmann*, Nonlinear Controllability and Optimal Control (1990)
134. *J.-P. Florens et al.*, Elements of Bayesian Statistics (1990)
135. *N. Shell*, Topological Fields and Near Valuations (1990)
136. *B. F. Doolin and C. F. Martin*, Introduction to Differential Geometry for Engineers (1990)
137. *S. S. Holland, Jr.*, Applied Analysis by the Hilbert Space Method (1990)
138. *J. Okniński*, Semigroup Algebras (1990)
139. *K. Zhu*, Operator Theory in Function Spaces (1990)
140. *G. B. Price*, An Introduction to Multicomplex Spaces and Functions (1991)
141. *R. B. Darst*, Introduction to Linear Programming (1991)
142. *P. L. Sachdev*, Nonlinear Ordinary Differential Equations and Their Applications (1991)
143. *T. Husain*, Orthogonal Schauder Bases (1991)
144. *J. Foran*, Fundamentals of Real Analysis (1991)
145. *W. C. Brown*, Matrices and Vector Spaces (1991)
146. *M. M. Rao and Z. D. Ren*, Theory of Orlicz Spaces (1991)
147. *J. S. Golan and T. Head*, Modules and the Structures of Rings (1991)
148. *C. Small*, Arithmetic of Finite Fields (1991)
149. *K. Yang*, Complex Algebraic Geometry (1991)
150. *D. G. Hoffman et al.*, Coding Theory (1991)
151. *M. O. González*, Classical Complex Analysis (1992)
152. *M. O. González*, Complex Analysis (1992)
153. *L. W. Baggett*, Functional Analysis (1992)
154. *M. Sniedovich*, Dynamic Programming (1992)
155. *R. P. Agarwal*, Difference Equations and Inequalities (1992)
156. *C. Brezinski*, Biorthogonality and Its Applications to Numerical Analysis (1992)
157. *C. Swartz*, An Introduction to Functional Analysis (1992)
158. *S. B. Nadler, Jr.*, Continuum Theory (1992)
159. *M. A. Al-Gwaiz*, Theory of Distributions (1992)
160. *E. Perry*, Geometry: Axiomatic Developments with Problem Solving (1992)
161. *E. Castillo and M. R. Ruiz-Cobo*, Functional Equations and Modelling in Science and Engineering (1992)
162. *A. J. Jerri*, Integral and Discrete Transforms with Applications and Error Analysis (1992)
163. *A. Charlier et al.*, Tensors and the Clifford Algebra (1992)
164. *P. Biler and T. Nadzieja*, Problems and Examples in Differential Equations (1992)
165. *E. Hansen*, Global Optimization Using Interval Analysis (1992)
166. *S. Guerre-Delabrière*, Classical Sequences in Banach Spaces (1992)

167. *Y. C. Wong*, Introductory Theory of Topological Vector Spaces (1992)
168. *S. H. Kulkarni and B. V. Limaye*, Real Function Algebras (1992)
169. *W. C. Brown*, Matrices Over Commutative Rings (1993)
170. *J. Loustau and M. Dillon*, Linear Geometry with Computer Graphics (1993)
171. *W. V. Petryshyn*, Approximation-Solvability of Nonlinear Functional and Differential Equations (1993)
172. *E. C. Young*, Vector and Tensor Analysis: Second Edition (1993)
173. *T. A. Bick*, Elementary Boundary Value Problems (1993)
174. *M. Pavel*, Fundamentals of Pattern Recognition: Second Edition (1993)
175. *S. A. Albeverio et al.*, Noncommutative Distributions (1993)
176. *W. Fulks*, Complex Variables (1993)
177. *M. M. Rao*, Conditional Measures and Applications (1993)
178. *A. Janicki and A. Weron*, Simulation and Chaotic Behavior of α-Stable Stochastic Processes (1994)
179. *P. Neittaanmäki and D. Tiba*, Optimal Control of Nonlinear Parabolic Systems (1994)
180. *J. Cronin*, Differential Equations: Introduction and Qualitative Theory, Second Edition (1994)
181. *S. Heikkilä and V. Lakshmikantham*, Monotone Iterative Techniques for Discontinuous Nonlinear Differential Equations (1994)
182. *X. Mao*, Exponential Stability of Stochastic Differential Equations (1994)
183. *B. S. Thomson*, Symmetric Properties of Real Functions (1994)
184. *J. E. Rubio*, Optimization and Nonstandard Analysis (1994)
185. *J. L. Bueso et al.*, Compatibility, Stability, and Sheaves (1995)
186. *A. N. Michel and K. Wang*, Qualitative Theory of Dynamical Systems (1995)
187. *M. R. Darnel*, Theory of Lattice-Ordered Groups (1995)
188. *Z. Naniewicz and P. D. Panagiotopoulos*, Mathematical Theory of Hemivariational Inequalities and Applications (1995)
189. *L. J. Corwin and R. H. Szczarba*, Calculus in Vector Spaces: Second Edition (1995)
190. *L. H. Erbe et al.*, Oscillation Theory for Functional Differential Equations (1995)
191. *S. Agaian et al.*, Binary Polynomial Transforms and Nonlinear Digital Filters (1995)
192. *M. I. Gil'*, Norm Estimations for Operation-Valued Functions and Applications (1995)
193. *P. A. Grillet*, Semigroups: An Introduction to the Structure Theory (1995)
194. *S. Kichenassamy*, Nonlinear Wave Equations (1996)
195. *V. F. Krotov*, Global Methods in Optimal Control Theory (1996)
196. *K. I. Beidar et al.*, Rings with Generalized Identities (1996)
197. *V. I. Arnautov et al.*, Introduction to the Theory of Topological Rings and Modules (1996)
198. *G. Sierksma*, Linear and Integer Programming (1996)
199. *R. Lasser*, Introduction to Fourier Series (1996)
200. *V. Sima*, Algorithms for Linear-Quadratic Optimization (1996)
201. *D. Redmond*, Number Theory (1996)
202. *J. K. Beem et al.*, Global Lorentzian Geometry: Second Edition (1996)
203. *M. Fontana et al.*, Prüfer Domains (1997)
204. *H. Tanabe*, Functional Analytic Methods for Partial Differential Equations (1997)
205. *C. Q. Zhang*, Integer Flows and Cycle Covers of Graphs (1997)
206. *E. Spiegel and C. J. O'Donnell*, Incidence Algebras (1997)
207. *B. Jakubczyk and W. Respondek*, Geometry of Feedback and Optimal Control (1998)
208. *T. W. Haynes et al.*, Fundamentals of Domination in Graphs (1998)
209. *T. W. Haynes et al.*, Domination in Graphs: Advanced Topics (1998)
210. *L. A. D'Alotto et al.*, A Unified Signal Algebra Approach to Two-Dimensional Parallel Digital Signal Processing

Additional Volumes in Preparation

A UNIFIED SIGNAL ALGEBRA APPROACH TO TWO-DIMENSIONAL PARALLEL DIGITAL SIGNAL PROCESSING

Louis A. D'Alotto
York College
The City University of New York
Queens, New York

Charles R. Giardina
College of Staten Island
The City University of New York
Staten Island, New York

Hua Luo
York College
The City University of New York
Queens, New York

MARCEL DEKKER, INC. NEW YORK • BASEL • HONG KONG

Library of Congress Cataloging-in-Publication Data

D'Alotto, Louis.
A unified signal algebra approach to two-dimensional parallel digital signal processing / Louis D'Alotto, Charles R. Giardina, Hua Luo.
p. cm.
Includes index.
ISBN 0-8247-0025-2
1. Signal processing—Digital techniques. 2. Algebras, Linear. I. Giardina, Charles R. (Charles Robert). II. Luo, Hua. III. Title.
TK5102.9.D34 1998
621.382'2'0285575—dc21

97-46946
CIP

The publisher offers discounts on this book when ordered in bulk quantities. For more information, write to Special Sales/Professional Marketing at the address below.

This book is printed on acid-free paper.

MARCEL DEKKER, INC.
270 Madison Avenue, New York, New York 10016
http://www.dekker.com

Current printing (last digit):
10 9 8 7 6 5 4 3 2 1

PRINTED IN THE UNITED STATES OF AMERICA

To our wives,
Zana, Betty and Qing

Preface

A Unified Signal Algebra Approach to Two-Dimensional Parallel Digital Signal Processing is a text dedicated to a modern approach in signal processing. A timely concern in the digital signal processing area is parallel processing. While numerous works have appeared on parallel computer architectures, relatively little information has been documented on the underlying principle for creating parallel digital signal processing algorithms to be used in these computers. This book fills the void by providing a unified signal algebra approach for creating parallel digital signal processing algorithms.

Parallel representations of basic signal processing operations occur naturally in the algebraic setting set forth herein. In the unified signal algebra approach, all operations are defined on digital signals, not on individual numbers that constitute the signals. Indeed, the arguments of all signal processing operations are functions represented by matrices – NOT NUMBERS. A matrix type representation, called a bound matrix, is used in describing digital signals. This representation, in the time domain, provides a useful alternative to the $\mathcal{Z}$-transform representation of digital signals. It captures all essential features of digital signals and provides necessary compression.

Operations on digital signals are specified using block diagrams. Together with bound matrices, the block diagram presentation provides a universal language for the description of important parallel digital signal processing algorithms. Throughout, an abundance of examples and execution type traces of block diagram algorithms are given. Many exercises are also presented. The exercises are intended to serve as an aid to understanding the material.

In terms of six fundamental operations, all other digital signal processing operations are formed under function composition. This is consistent with reduced instruction set computer philosophy, whereby a "small number of building blocks" can be used to form useful operations. Each of the six fundamental operations is induced by similar operations in the domain or range space and it is this technique that provides the unified algebraic methodology. While the methodology contained herein is useful in various signal processing environments, this text is self-contained and always remains at a "how to" level.

Two dimensional digital signal processing is mainly concerned with the extraction of information from signals. The steps involved in the information extraction procedure include some or all of the following:

1.) Digital signal creation
2.) Digital signal to digital signal operation
3.) Digital signal to real number operation
4.) Real number to decision operation

In this book, we concentrate on step 2 above. Essentially, the primary and only difference between two dimensional and single dimensional signal processing is greater computational complexity when dealing with the higher dimensional signals.

Even though various digital signal processing techniques are independent of dimension, the representation of such signals is dimensionally dependent. Matrix type notation, called bound matrices, is used to represent two dimensional digital signals. This structure is introduced in the first chapter and employed throughout.

All parallel algorithms are specified using block diagrams. This allows easily describable traces and walk-throughs of all algorithms and is consistent with data flow procedural descriptions.

Using two dimensional signals allows for a rich processing environment. This environment is exploited by using two dimensional techniques for efficient processing of single dimensional digital signals. Indeed, the last chapter of the text is dedicated to this type of optimization.

Two dimensional wraparound signal processing is also presented in the text. A self-contained chapter describes the algebraic environment necessary for processing these types of signals. Again, bound matrix representation is introduced and block diagram specification of all algorithms is given.

This book is intended to be used as a text in digital signal processing courses for graduate students. It also provides a useful reference for applied mathematicians, computer scientists, and electrical engineers who are interested in algebraic techniques for providing parallel algorithms.

Louis A. D'Alotto
Charles R. Giardina
Hua Luo

Contents

Preface v

1 Two Dimensional Signals **1**
1.1 Two Dimensional Digital Signals 1
1.2 Important Two Dimensional Digital Signals 5
1.3 Bound Matrix Representation 8
1.4 Exercises 14

2 Fundamental Operations on Two Dimensional Signals **17**
2.1 Fundamental Range Induced Operations 17
2.2 Terms Involving Fundamental Range Induced Operations 20
2.3 Additional Important Digital Signals and Macro Operators 26
2.4 Equational Identities Involving Range Induced Operations 29
2.5 Fundamental Domain Induced Operations 35
2.6 Terms Involving Fundamental Domain Induced Operations 42
2.7 Further Equational Identities 50
2.8 Exercises 53

3 Convolution of Digital Signals **59**
3.1 The Support Region for Convolution 59
3.2 Convolution for Signals of Finite Support 61
3.3 Bound Matrices for Convolution 65
3.4 Parallel Convolution Algorithms 68
3.5 Convolution of Signals of Non-Finite Support 73
3.6 Banach Algebra Properties of Convolution in l_1 79
3.7 Filtering by Convolution 83
3.8 Correlation 92

3.9 Applications of Correlation 103
3.10 Exercises 109

4 $\mathcal{Z}$ Transforms **115**
4.1 Formal Introduction to $\mathcal{Z}$ Transforms 115
4.2 Some Operations Involving $\mathcal{Z}$ Transforms 118
4.3 $\mathcal{Z}$ Transforms for Digital Signals of Finite Support . . 125
4.4 Laurent Expansions 129
4.5 $\mathcal{Z}$ Transform for l_1 Signals and Signals of Arbitrary Support 143
4.6 The $\mathcal{Z}$ Transform for Calculating Correlation 146
4.7 Transfer Functions 148
4.8 Exercises: 153

5 Difference Equations **159**
5.1 Function Equations 159
5.2 Linear Space Invariant Difference Equations 161
5.3 Difference Equations Involving One Unit Translation . 168
5.4 $\mathcal{Z}$ Transforms for Solving Difference Equations 178
5.5 Exercises 185

6 Wraparound Signal Processing **187**
6.1 Wraparound Signals 187
6.2 Range Induced Operations and Terms for Wraparound Signals 190
6.3 Domain Induced Operations for Wraparound Signals . 195
6.4 Set Morphology in $\mathbf{Z_n} \times \mathbf{Z_m}$ 204
6.5 Sequential and Parallel Formulation of Wraparound Convolution 212
6.6 Sequential and Parallel Formulations of Wraparound Correlation 218
6.7 Wraparound $\mathcal{Z}$ Transforms 223
6.8 Discrete Fourier Transforms 230
6.9 Wraparound Algebra in $\mathbf{Z}_n \times \mathbf{Z}_m$ 234
6.10 Exercises 243

7 Parallel Multidimensional Algorithms for Single Dimensional Signal Processing 249

7.1 Two Dimensional Processing of One Dimensional Signals . . . 249
7.2 Diagonal Transform from One Dimension into Two Dimensions . . . 252
7.3 Kronecker Products . . . 254
7.4 Triple Convolution . . . 259
7.5 Triple Convolution Performed Using Parallel Two Dimensional Convolution . . . 261
7.6 Volterra Convolution and Volterra Series . . . 265
7.7 Parallel Algorithms for Volterra Convolution . . . 269
7.8 Exercises . . . 275

Appendix 279

References 287

1

Two Dimensional Signals

1.1 Two Dimensional Digital Signals

For the most part, in the text we will be presenting operations involving two dimensional digital signals f. A *digital signal* is a function which provides the values of some attribute associated with some object, system or physical phenomena. The set of all two dimensional digital signals is denoted by $\Re^{Z \times Z}$, which is the set of all real valued functions defined at every pair of integers. So $f \in \Re^{Z \times Z}$ means

$$f : \mathbf{Z} \times \mathbf{Z} \rightarrow \Re$$

In other words, a digital signal has domain consisting of all the lattice points (n, m) where m and n are in $\mathbf{Z}$. Moreover, their codomain is the set of real values, whereas the range of any specific digital signal is the set of real values which the signal actually attains. The following demonstrates an example of a two dimensional digital signal.

Example 1.1 *Consider the two dimensional digital signal f where*

$$f(n, m) = \begin{cases} 4 & at\,(-1, -1) \\ 3 & at\,(0, 1) \\ -2 & at\,(1, 0) \\ 0 & elsewhere \end{cases}$$

Observe that the range of f is the set

$$\{-2,\ 0,\ 3,\ 4\}$$

Notice that the two dimensional digital signal f, given in Example 1.1, is illustrated in Figure 1.1.

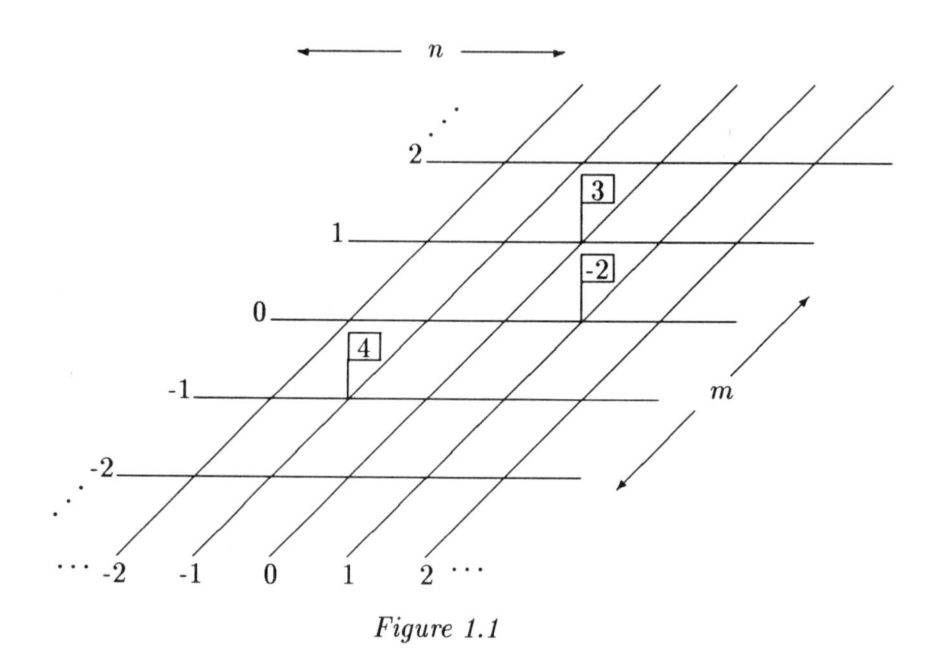

Figure 1.1

The lattice points $\mathbf{Z} \times \mathbf{Z}$ are those points formed by the intersection of a horizontal and vertical line. At every lattice point (n, m) for which f attains a nonzero value, a "flag" is used in prescribing the value $f(n, m)$. At any point for which there is no flag, it is assumed that f is zero valued.

Example 1.2 *Suppose that the two dimensional digital signal f is given where*

$$f(n,m) = | n - m |$$

Then

$$f(-1,1) = | -1 - 1 | = 2$$

and

$$f(2,-1) = | 2 + 1 | = 3$$

also

$$f(n,m) = 0 \quad \text{for } n = m$$

A graphical illustration of the signal f given in Example 1.2 is provided in Figure 1.2. A few points at which the function attains nonzero values are given in this figure. An infinite number of flags should be employed in this diagram and the dots are used to indicate that this is the case.

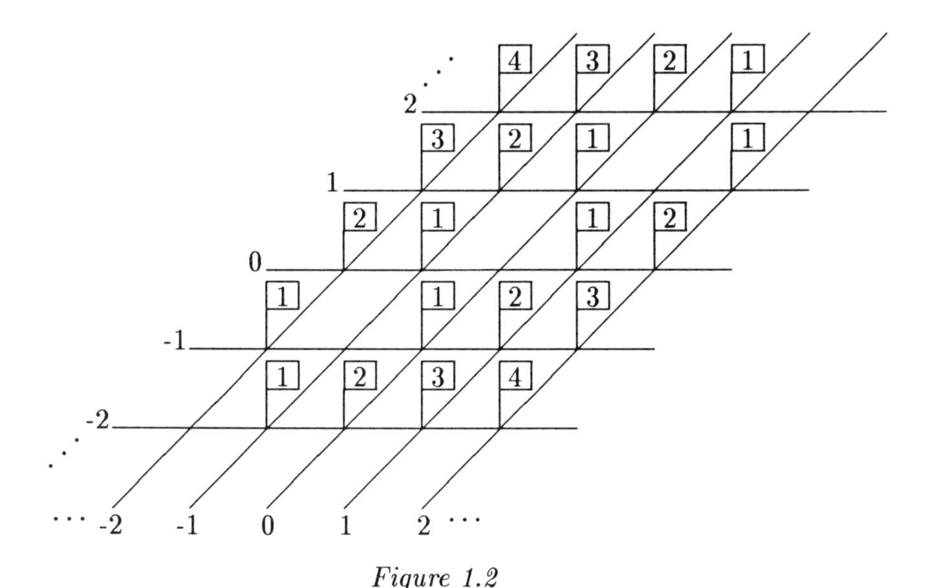

Figure 1.2

As previously mentioned, signals most often provide values associated with some physical phenomena. Accordingly, signals are frequently obtained from sensor readings. For mapping purposes only, we assume that the Earth is flat and infinite. The sensor, for instance, might be an accelerometer. An accelerometer (held vertically) which is not moving relative to the Earth will measure vertical acceleration. Most importantly, the reading will vary in a known manner depending on whether it is held on a mountain top or valley bottom. Intuitively, in the former case, the accelerometer reading will be less than the reading obtained in the latter case. Accordingly, this instrument can be employed to measure variation in vertical heights. We will now use it for this purpose.

Our objective is to find a two dimensional digital signal representative of the mountain region in Figure 1.3(a). We will assume that a coordinate system is located with its origin on top of the mountain. At every integer unit along the east and north directions, a reading shall be taken using the accelerometer and recorded on a flag. This is the sampling procedure. The integer unit may be every 10 feet, it could be 100 feet, or 10 yards. The smaller this quantity is, the more accurate the terrain map. Figure 1.3(b) provides a two dimensional digital signal representation of the mountain. Note that some readings, in Figure 1.3(b), are taken behind the mountain and are therefore not visible in Figure 1.3(a).

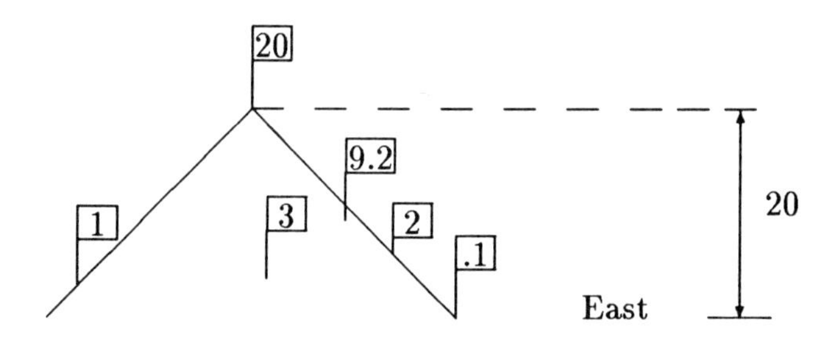

Figure 1.3(a)
Side View of Mountain

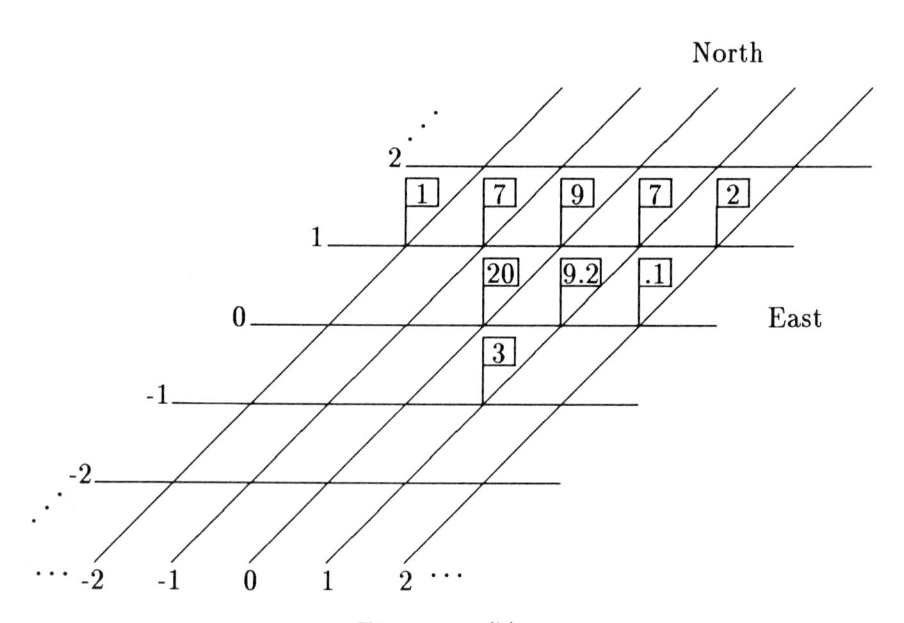

Figure 1.3(b)
Top View of Mountain, Using a Two Dimensional Digital Signal.

Two dimensional digital signals are defined at every integer pair (n, m). Often, however, due to lack of information, there may be points at which values of a signal are not observed. At these points we define the signal to have value 0 there. A quite different situation holds in Chapter 6 where two dimensional wraparound signals are defined and operated upon. These signals are defined only in some rectangular subset of the lattice points.

1.2 Important Two Dimensional Digital Signals

Several important signals in $\Re^{Z\times Z}$ are given next. In each case, a single real value is assigned to every integer. Several of these functions are illustrated in Figures 1.4(a) through 1.4(e). For illustrative purposes we only display a few function value flags.

Among the simplest functions in $\Re^{Z\times Z}$ are the ones which are of constant value at every integer. These functions are denoted by $a_{Z\times Z}$ and so

$$a_{Z\times Z}(n, m) = a$$

For instance, the function $3_{Z\times Z}$ is illustrated in figure 1.4(a). Of special interest is the zero function $0_{Z\times Z}$, and the unity function $1_{Z\times Z}$ which have values zero and one, respectively at each integer. These two functions are illustrated in Figures 1.4(b) and 1.4(c), respectively. The delta, or impulse function δ is a function in $\Re^{Z\times Z}$ which has value 1 at the origin and is zero elsewhere in $\mathbf{Z}\times\mathbf{Z}$. Thus,

$$\delta(n, m) = \begin{cases} 1 & (n, m) = (0, 0) \\ 0 & otherwise \end{cases}$$

It is illustrated in Figure 1.4(d).

The first quadrant unit step function, denoted by u, is also an important digital signal. It attains the value 1 for lattice points consisting of nonnegative integer pairs, and it is of value 0 otherwise, hence

$$u(n, m) = \begin{cases} 1 & n \in \{0, 1, 2, ...\} \text{ and } m \in \{0, 1, 2, ...\} \\ 0 & otherwise \end{cases}$$

This function is illustrated in Figure 1.4(e).

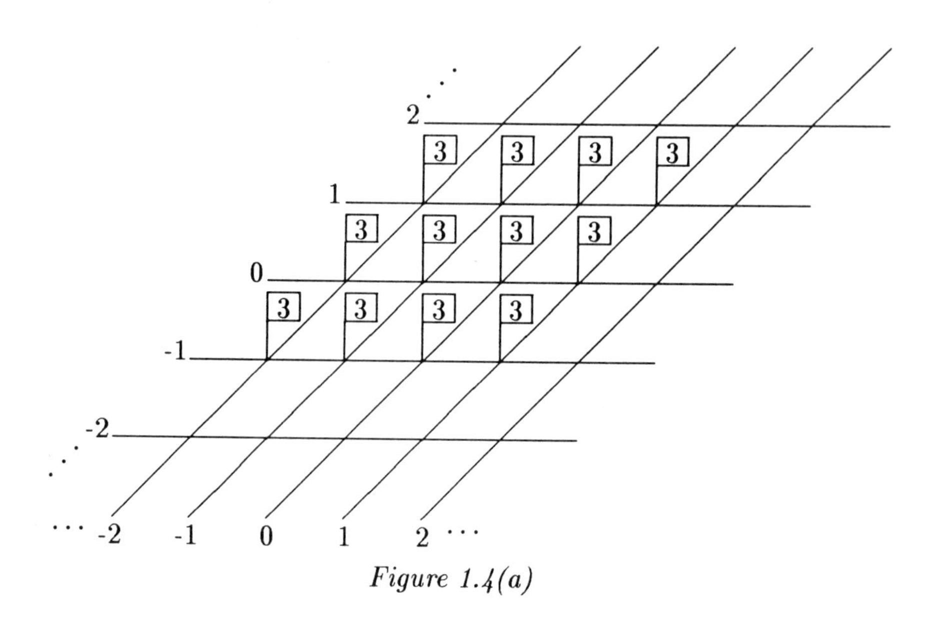

Figure 1.4(a)

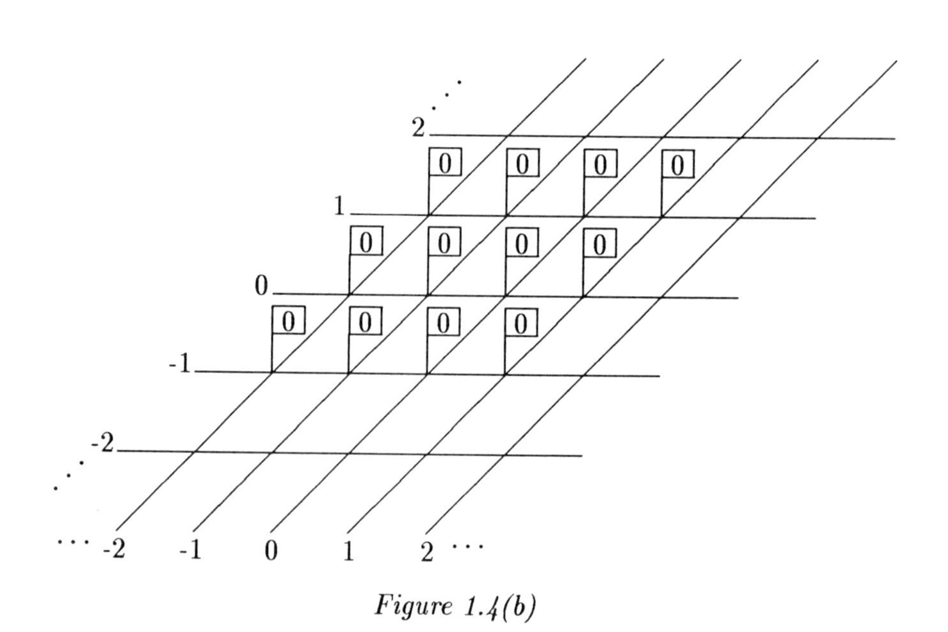

Figure 1.4(b)

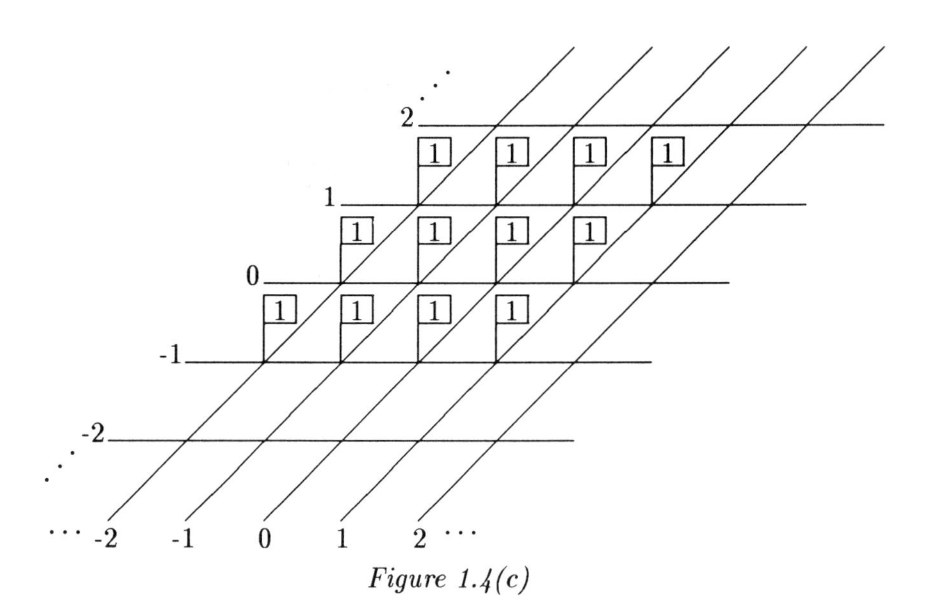

Figure 1.4(c)

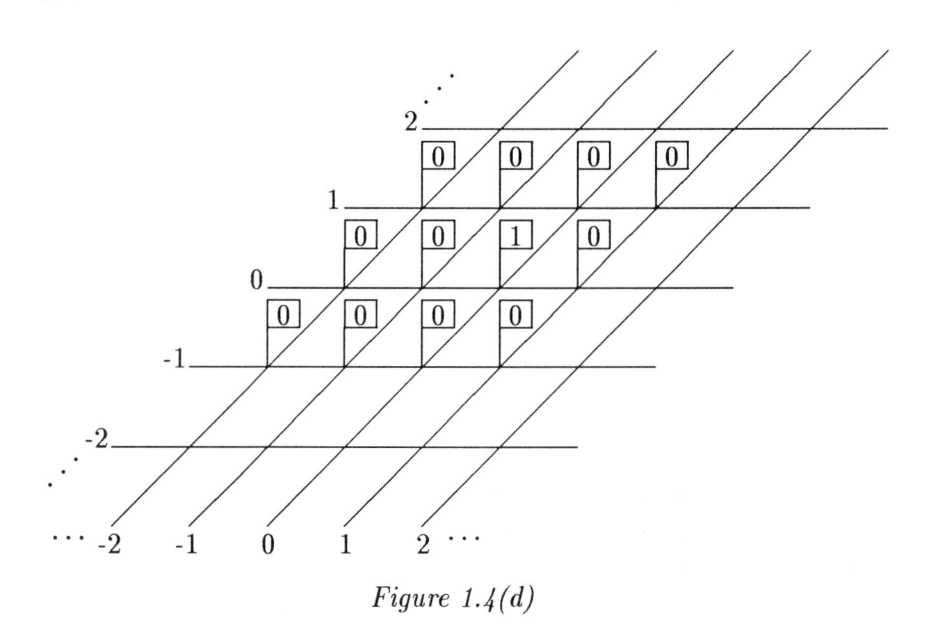

Figure 1.4(d)

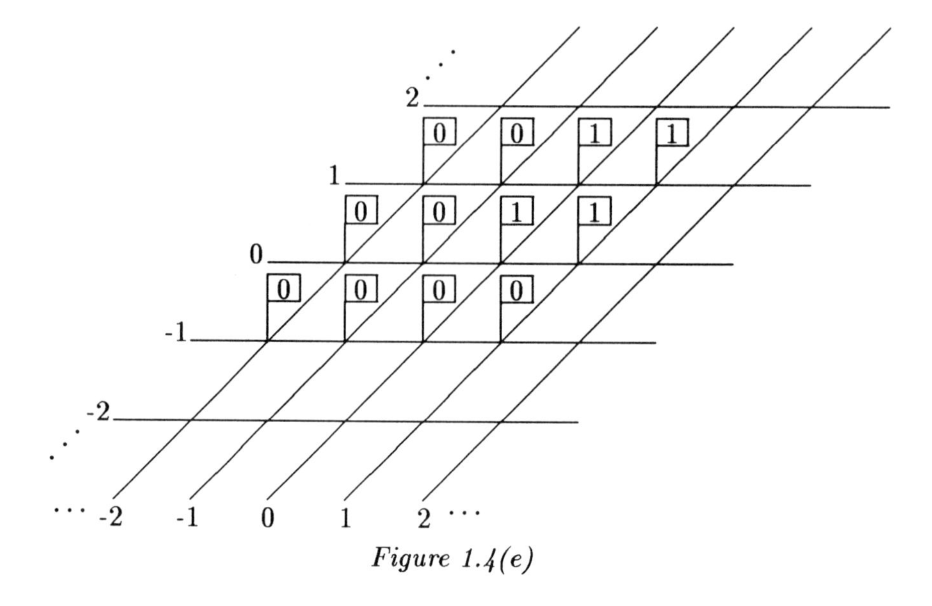

Figure 1.4(e)

1.3 Bound Matrix Representation

For two dimensional signals, f will be represented using bound matrices. In this representation, a finite or infinite matrix is employed consisting of values in the range of f. Specifically, we write f as

$$\begin{pmatrix} & \vdots & \vdots & \vdots & \\ \cdots & f(-1,1) & f(0,1) & f(1,1) & \cdots \\ \cdots & f(-1,0) & \boxed{f(0,0)} & f(1,0) & \cdots \\ \cdots & f(-1,-1) & f(0,-1) & f(1,-1) & \cdots \\ & \vdots & \vdots & \vdots & \end{pmatrix}$$

The three dots on the left, right, above and below the elements denote values which exist but are omitted in the representation. A box is placed around the value of f at the origin. Other locations of values of f are found by their relative position in the lattice grid $\mathbf{Z} \times \mathbf{Z}$.

Example 1.3 *Some of the functions illustrated in Figures* 1.4(a) *through* 1.4(e) *will be represented as bound matrices. The constant function in Figure* 1.4(a) *is represented as*

$$a_{Z\times Z} = \begin{pmatrix} & \cdot & \cdot & \cdot & \\ & \cdot & \cdot & \cdot & \\ & \cdot & \cdot & \cdot & \\ \cdots & a & a & a & \cdots \\ \cdots & a & \boxed{a} & a & \cdots \\ \cdots & a & a & a & \cdots \\ & \cdot & \cdot & \cdot & \\ & \cdot & \cdot & \cdot & \\ & \cdot & \cdot & \cdot & \end{pmatrix}$$

With $a = 3$,

$$3_{Z\times Z} = \begin{pmatrix} & \cdot & \cdot & \cdot & \\ & \cdot & \cdot & \cdot & \\ & \cdot & \cdot & \cdot & \\ \cdots & 3 & 3 & 3 & \cdots \\ \cdots & 3 & \boxed{3} & 3 & \cdots \\ \cdots & 3 & 3 & 3 & \cdots \\ & \cdot & \cdot & \cdot & \\ & \cdot & \cdot & \cdot & \\ & \cdot & \cdot & \cdot & \end{pmatrix}$$

The delta function in Figure 1.4(d) *is represented as*

$$\delta = \begin{pmatrix} & \cdot & \cdot & \cdot & \\ & \cdot & \cdot & \cdot & \\ & \cdot & \cdot & \cdot & \\ \cdots & 0 & 0 & 0 & \cdots \\ \cdots & 0 & \boxed{1} & 0 & \cdots \\ \cdots & 0 & 0 & 0 & \cdots \\ & \cdot & \cdot & \cdot & \\ & \cdot & \cdot & \cdot & \\ & \cdot & \cdot & \cdot & \end{pmatrix}$$

Finally, the first quadrant unit step function, illustrated in Figure 1.4(e) is given by

$$u = \begin{pmatrix} & \cdot & \cdot & \cdot & \\ & \cdot & \cdot & \cdot & \\ & \cdot & \cdot & \cdot & \\ \cdots & 0 & 1 & 1 & \cdots \\ \cdots & 0 & \boxed{1} & 1 & \cdots \\ \cdots & 0 & 0 & 0 & \cdots \\ & \cdot & \cdot & \cdot & \\ & \cdot & \cdot & \cdot & \\ & \cdot & \cdot & \cdot & \end{pmatrix}$$

More often than not, signals in $\Re^{Z \times Z}$ will be zero outside the first quadrant. Signals possessing this property are said to have support in the first quadrant set. We will let S denote the subset of $\Re^{Z \times Z}$ containing these signals. Signals in this class can be represented in a more simple manner than previously described. We will write

$$f = \begin{pmatrix} \cdot & \cdot & \cdot \\ \cdot & \cdot & \cdot \\ \cdot & \cdot & \cdot \\ f(0,2) & f(1,2) & \cdots \\ f(0,1) & f(1,1) & \cdots \\ \boxed{f(0,0)} & f(1,0) & \cdots \end{pmatrix}^{0}$$

The zero in the upper right, outside the right parenthesis, means that at all lattice points not in the first quadrant, f has value 0.

A *support region* of a function is a subset of $\mathbf{Z} \times \mathbf{Z}$ outside of which f has value zero. A support region of f is denoted by $\text{supp}(f)$ and must contain all points at which f is not zero. It can also contain points in the domain at which f does equal zero. Accordingly, $\text{supp}(f)$ is often not unique.

Of particular interest is the set F of all two dimensional digital signals with finite support. A function f is in the class F if and only if the cardinality of some support region is finite, that is

$$\text{card}(\text{supp}(f)) < \infty.$$

Any two dimensional signal in F can be represented using the more compact, n rows by m columns, bound matrix representation

$$f = \begin{pmatrix} a_{11} & a_{12} & \cdots & a_{1m} \\ a_{21} & a_{22} & \cdots & a_{2m} \\ \cdot & \cdot & & \cdot \\ \cdot & \cdot & & \cdot \\ \cdot & \cdot & & \cdot \\ a_{n1} & a_{n2} & \cdots & a_{nm} \end{pmatrix}_{r,t}^{0}$$

Here, $f(r,t) = a_{11}$ and all other values of f are either 0 or given by the values a_{ij} in the matrix. Corresponding lattice points for these values a_{ij} are obtained by situating the matrix on the integral lattice.

Example 1.4 *Observe that the signal*

$$f(n,m) = \begin{cases} 4 & at\ (-1,-1) \\ 3 & at\ (0,1) \\ -2 & at\ (1,0) \\ 0 & elsewhere \end{cases}$$

illustrated in Figure 1.1 *is of finite support, thus not only can we represent it as*

$$f = \begin{pmatrix} & \cdot & \cdot & \cdot & \cdot & \cdot & \cdot & \cdot & \\ & \cdot & \cdot & \cdot & \cdot & \cdot & \cdot & \cdot & \\ & \cdot & \cdot & \cdot & \cdot & \cdot & \cdot & \cdot & \\ \cdots & 0 & 0 & 0 & 0 & 0 & 0 & 0 & \cdots \\ \cdots & 0 & 0 & 0 & 3 & 0 & 0 & 0 & \cdots \\ \cdots & 0 & 0 & 0 & \boxed{0} & -2 & 0 & 0 & \cdots \\ \cdots & 0 & 0 & 4 & 0 & 0 & 0 & 0 & \\ & \cdot & \cdot & \cdot & \cdot & \cdot & \cdot & \cdot & \\ & \cdot & \cdot & \cdot & \cdot & \cdot & \cdot & \cdot & \\ & \cdot & \cdot & \cdot & \cdot & \cdot & \cdot & \cdot & \end{pmatrix}$$

but we can represent it in the more compressed form

$$f = \begin{pmatrix} 0 & 3 & 0 \\ 0 & 0 & -2 \\ 4 & 0 & 0 \end{pmatrix}_{-1,1}^{0}$$

In the previous example we saw the more compressed form of representing two dimensional digital signals of finite support. For these signals we displayed a "window" of values which included all the places and corresponding values where the signal is nonzero. Of course, such a representation is not unique. An infinite number of bound matrices can represent the same signal. However, for any nonzero signal of finite support, a unique minimal representation can be obtained. It is called the *minimal bound matrix.* This representation always possesses a nonzero value on each "boundary" point of the matrix. That is, it has a nonzero element on the leftmost and rightmost column as well as a nonzero value on the uppermost, and lowermost rows. The second matrix representation for f, given in the last example, is minimal.

Example 1.5 *The only functions of finite support given in Figures* 1.4(a) *through* 1.4(e) *are* $0_{Z\times Z}$ *and* δ. *Any support region of* δ *must contain the origin, and any bound matrix representation must have a value* 1 *at the origin. For instance*

$$\delta = \begin{pmatrix} 1 & 0 & 0 \\ 0 & 0 & 0 \end{pmatrix}^{0}_{0,0}$$

and

$$\delta = \begin{pmatrix} 0 & 0 \\ 0 & 1 \end{pmatrix}^{0}_{-1,1}$$

whereas,

$$\delta = (1)^{0}_{0,0}$$

is a minimal bound matrix representation of δ.

We will now give the conversion formula between a given bound matrix representation and function type representation. This is given mainly for computer implementation purposes.

Let f be a digital signal with a rectangular support region consisting of n rows and m columns of lattice points, so

$$\text{supp}(f) = \{(i,j) \text{ where } p \leq i \leq p+m-1,\ q-n+1 \leq j \leq q\}$$

Thus, we are guaranteed that $f(i,j) = 0$ for all (i,j) outside supp(f) and f can be represented with the n row by m column bound matrix

$$\begin{pmatrix} a_{11} & a_{12} & \dots & a_{1m} \\ a_{21} & a_{22} & \dots & a_{2m} \\ \cdot & \cdot & & \cdot \\ \cdot & \cdot & & \cdot \\ \cdot & \cdot & & \cdot \\ a_{n1} & a_{n2} & \dots & a_{nm} \end{pmatrix}_{p,q}^{0}$$

Given a_{rt} we can find the corresponding lattice point at which this value is attained, indeed,

$$a_{rt} = f(p+t-1, q-r+1)$$

For instance,

$$a_{11} = f(p,q)$$

$$a_{12} = f(p+1,q)$$

$$a_{1m} = f(p+m-1,q)$$

$$a_{21} = f(p,q-1)$$

$$a_{n1} = f(p,q-n+1)$$

$$a_{nm} = f(p+m-1,q-n+1)$$

Going the other way, if $f(i,j)$ is given for (i,j) in supp(f) then

$$f(i,j) = a_{i-p+1,\,q-j+1}$$

We have seen that for every function f in F, there exists an equivalence class of bound matrices corresponding to f. The minimal bound matrix corresponding to f ($f \neq 0_{Z\times Z}$) may be thought of as the most ideal representation. The minimal bound matrix is also called a coset leader. However, for clarity purposes, nonminimal representation is often helpful. On occasion we will use other bound matrix representations besides the minimal one. This will be seen as we process bound matrices using addition, multiplication, convolution, and other operations.

1.4 Exercises

1. Which of the following functions, are two dimensional digital signals:

$$a) \quad f(n,m) = | n \cdot m | \quad n,m \in \mathbf{Z}$$

$$b) \quad f(n,m) = n \cdot m \quad n,m \in \mathbf{Z}$$

$$c) \quad f(n,m) = \frac{1}{nm^{\frac{1}{3}}} \quad n,m \in \mathbf{Z}$$

$$d) \quad f(n,m) = \sqrt{nm} \quad n,m \in \mathbf{Z}$$

2. Not all of the following functions are in $\Re^{Z \times Z}$. Identify those which are, and illustrate each by providing a diagram.

$$a) \quad f(n,m) = | n |^{\frac{1}{2}} m^{\frac{1}{3}} \quad for\ n,m \in \mathbf{Z}$$

$$b) \quad f(n,m) = \cos(\tfrac{n\pi}{2}) \quad for\ n \in \mathbf{Z}$$

$$c) \quad f(x,y) = \begin{cases} 1 & (x,y) \text{ a lattice point in } \Re \times \Re \\ 0 & (x,y) \text{ not a lattice point in } \Re \times \Re \end{cases}$$

3. For each of the functions given in Exercise 1, find both the domain and range sets.

4. Find the domain and range sets for the functions given in Exercise 2.

5. Provide a bound matrix representation for the two dimensional digital signal f, where

$$f(n,m) = \begin{cases} 0 & n \le 0\ and\ all\ m \\ 1 & n > 0\ and\ m \ge 0 \\ 0 & n > 0\ and\ m < 0 \end{cases}$$

Explain why f is not the unit step function u, defined in this chapter.

6. Is the signal f of Exercise 5 of finite support?

7. Let

$$f = \begin{pmatrix} 3 & 4 \\ 2 & -1 \end{pmatrix}_{2,4}^{0}$$

a) Explain why f is of finite support.

b) Is it correct to write

$$f = \begin{pmatrix} 0 & 0 & 0 \\ 0 & 3 & 4 \\ 0 & 2 & -1 \end{pmatrix}_{1,5}^{0}$$

c) Show that one of the support regions of f is given by

$$\text{supp}(f) = \{(2,4),(3,4),(-1,4),(2,3),(3,3)\}$$

d) Are there any smaller (in cardinality) support regions for f? If so find one.

8. Is there a two dimensional digital signal f, for which any subset of the lattice points is a support region? If so, find one such signal.

9. Is there a two dimensional digital signal for which no subset of the lattice points is a support region? If so, find one such signal.

10. Show that there is a one to one correspondence between the set of all two dimensional signals of finite support, excluding $0_{Z\times Z}$, and the set of all minimal bound matrices which represent these signals.

2

Fundamental Operations on Two Dimensional Signals

2.1 Fundamental Range Induced Operations

Throughout the rest of the text, we will write, "digital signals" in place of "two dimensional digital signals". Several operations can be defined on these signals which are analogous to operations involving the real number system. This is because the range space of any digital signal is a subset of the reals. Accordingly, the operations defined herein are called *range induced operations.* These operations are defined in a pointwise manner; they are performed coordinatewise at each lattice point using similar operations involving real numbers. Precisely three fundamental range induced operations will be defined.

Perhaps the simplest operation on digital signals is the binary addition operation. It will be denoted by ADD. This operation has two signals as input and yields one signal as output. Hence,

$$ADD : \Re^{Z\times Z} \times \Re^{Z\times Z} \rightarrow \Re^{Z\times Z}$$

and it is defined by using the rule

$$ADD(f,g)(n,m) = f(n,m) + g(n,m)$$

Thus, to obtain the sum or addition of two bound matrices, f and g, we add their values pointwise. This operation is also illustrated in a graphical manner using block diagrams, as follows:

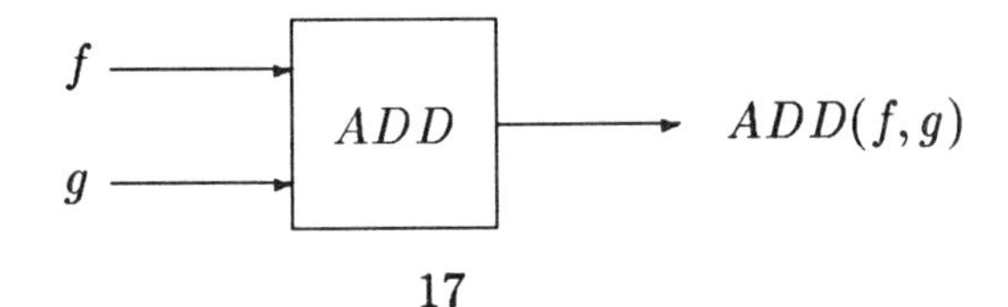

When no confusion should result, instead of writing the addition operation using the prefix notation $ADD(f,g)$, we use the more customary infix notation $f+g$. The two symbolisms shall be used interchangeably henceforth.

Example 2.1 *Suppose that the addition of*

$$f = \begin{pmatrix} 2 & 3 \\ 1 & 4 \\ -1 & 1 \end{pmatrix}^{0}_{3,4}$$

and

$$g = \begin{pmatrix} 2 & -1 & 3 \\ 1 & 0 & 2 \end{pmatrix}^{0}_{3,3}$$

is desired. In this case, it is more useful not to employ minimal matrices to represent f and g. Indeed, we write

$$f = \begin{pmatrix} 2 & 3 & 0 \\ 1 & 4 & 0 \\ -1 & 1 & 0 \end{pmatrix}^{0}_{3,4}$$

and

$$g = \begin{pmatrix} 0 & 0 & 0 \\ 2 & -1 & 3 \\ 1 & 0 & 2 \end{pmatrix}^{0}_{3,4}$$

Next, we add pointwise to obtain

$$f+g = \begin{pmatrix} 2 & 3 & 0 \\ 3 & 3 & 3 \\ 0 & 1 & 2 \end{pmatrix}^{0}_{3,4}$$

For instance, notice that

$$(f+g)(4,3) = f(4,3) + g(4,3) = 4 - 1 = 3$$

Two other fundamental operations which are structurally similar to addition are the multiplication operation, denoted by $MULT$, and the maximum operation, denoted by MAX. Both of these operations are binary, thereby taking two signals as input arguments and yielding a signal as output. Thus,

$$MULT : \Re^{Z\times Z} \times \Re^{Z\times Z} \rightarrow \Re^{Z\times Z}$$

and

$$MAX : \Re^{Z\times Z} \times \Re^{Z\times Z} \rightarrow \Re^{Z\times Z}$$

They are defined using the pointwise definition. Indeed,

$$MULT(f,g)(n,m) = f(n,m) \cdot g(n,m)$$

and

$$MAX(f,g)(n,m) = \max(f(n,m), g(n,m))$$

Again, we will abuse notation by letting

$$MULT(f,g) = f \cdot g$$

Additionally, we will often write $f \vee g$ instead of $MAX(f,g)$.

The block diagrams used in specifying the multiplication and maximum operations are given respectively as

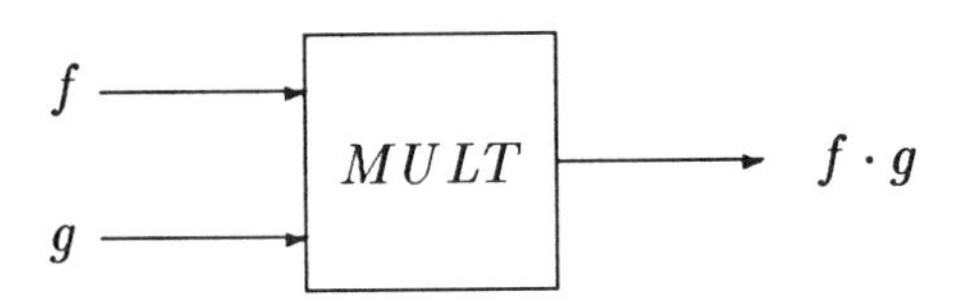

and

f → MAX → $f \vee g$
g →

Example 2.2 *Consider again, the functions f and g from Example 2.1. Since*

$$f = \begin{pmatrix} 2 & 3 & 0 \\ 1 & 4 & 0 \\ -1 & 1 & 0 \end{pmatrix}_{3,4}^{0}$$

and

$$g = \begin{pmatrix} 0 & 0 & 0 \\ 2 & -1 & 3 \\ 1 & 0 & 2 \end{pmatrix}_{3,4}^{0}$$

then

$$f \cdot g = \begin{pmatrix} 0 & 0 & 0 \\ 2 & -4 & 0 \\ -1 & 0 & 0 \end{pmatrix}_{3,4}^{0}$$

and

$$f \vee g = \begin{pmatrix} 2 & 3 & 0 \\ 2 & 4 & 3 \\ 1 & 1 & 2 \end{pmatrix}_{3,4}^{0}$$

2.2 Terms Involving Fundamental Range Induced Operations

We will now apply the three fundamental operations, given in the last section, in succession to various signals. Hence, we will use the fundamental range induced operations along with function composition. The resulting operations are called *terms* or *macro* operations.

One of the most simple macro operations is the unary square law device, denoted SQ. We will also write f^2 instead of $SQ(f)$. The given operation is defined by

$$SQ : \Re^{Z \times Z} \to \Re^{Z \times Z}$$

where

$$SQ(f)(n,m) = f^2(n,m)$$

The block diagram illustrating this operation is

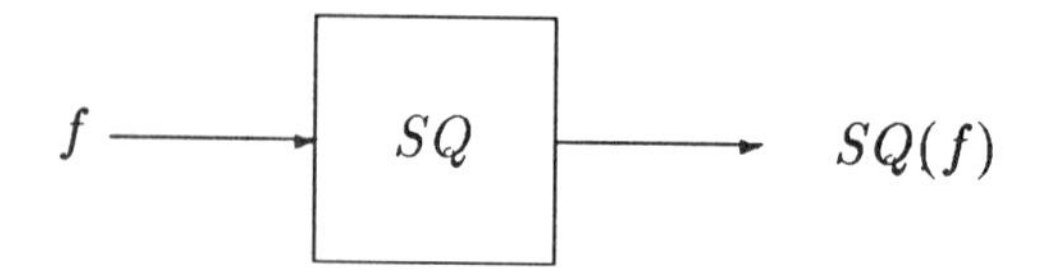

The reason why SQ is a term, is that the output of

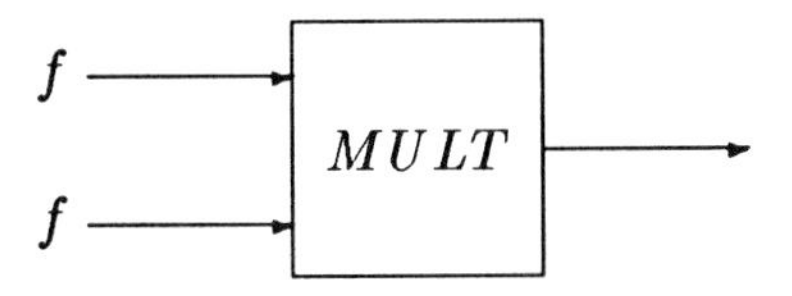

is also $SQ(f)$. That is, the square law device is nothing more than the multiplication of a signal with itself.

Example 2.3 *If*

$$f = \begin{pmatrix} 2 & 3 \\ 1 & 4 \\ -1 & 1 \end{pmatrix}^{0}_{3,4}$$

we have

$$f \longrightarrow \boxed{SQ} \longrightarrow f^2 = \begin{pmatrix} 4 & 9 \\ 1 & 16 \\ 1 & 1 \end{pmatrix}^{0}_{3,4}$$

The negation of a signal f is denoted by $MINUS(f)$. It will also be denoted by $-f$. In any case, it is a unary operation which is defined by

$$MINUS : \Re^{Z\times Z} \rightarrow \Re^{Z\times Z}$$

where

$$MINUS(f)(n,m) = -f(n,m)$$

Using the fact that the following two block diagrams always provide the same output shows that $MINUS$ is a term.

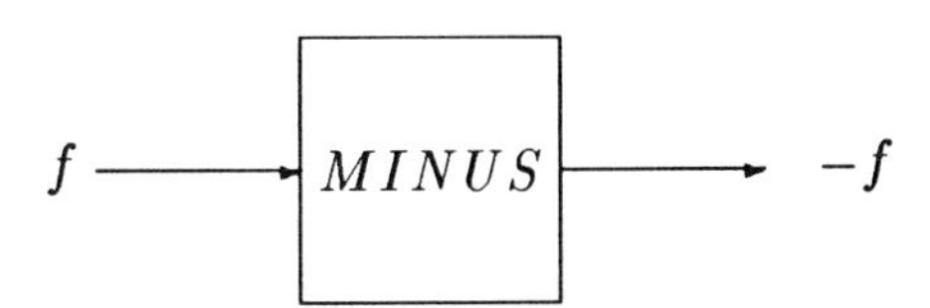

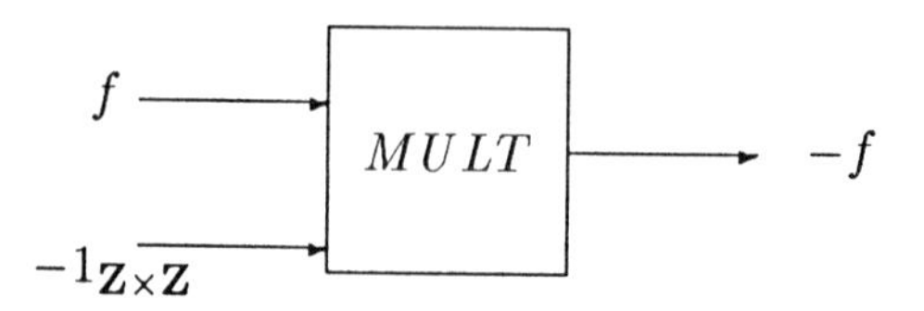

The last block diagram involved the multiplication of the two input signals , f, and the constant function $-1_{Z\times Z}$ of value -1 everywhere in the integral lattice.

Example 2.4 *Let*

$$f = \left(\begin{array}{rr} 2 & 3 \\ 1 & 4 \\ -1 & 1 \end{array}\right)^{0}_{3,4}$$

then

$$-f = \left(\begin{array}{rr} -2 & -3 \\ -1 & -4 \\ 1 & -1 \end{array}\right)^{0}_{3,4}$$

As described above, we can find $-f$ using the latter block diagram. In this case, it is useful to write $-1_{Z\times Z}$ using the more compact bound matrix representation

$$\left(\begin{array}{rr} -1 & -1 \\ -1 & -1 \\ -1 & -1 \end{array}\right)^{-1}_{3,4}$$

Recall, the -1 outside the right parenthesis means that the function given by this bound matrix is of value -1 for all lattice points outside of $\{(3,4),(4,4),(3,3),(4,3),(3,2),(4,2)\}$. Tracing through the block diagram we have

$$\left(\begin{array}{rr} 2 & 3 \\ 1 & 4 \\ -1 & 1 \end{array}\right)^{0}_{3,4}, \quad \left(\begin{array}{rr} -1 & -1 \\ -1 & -1 \\ -1 & -1 \end{array}\right)^{-1}_{3,4} \longrightarrow \boxed{MULT} \longrightarrow \left(\begin{array}{rr} -2 & -3 \\ -1 & -4 \\ 1 & -1 \end{array}\right)^{0}_{3,4}$$

Another important term is the binary minimum operator, denoted by MIN, which is defined by

$$MIN : \Re^{Z\times Z} \times \Re^{Z\times Z} \rightarrow \Re^{Z\times Z}$$

where

$$MIN(f,g)(n,m) = \min(f(n,m), g(n,m))$$

The minimum operation on f and g is also denoted by $f \wedge g$. MIN is a term because

$$f \wedge g = -(-f \vee -g)$$

Using block diagrams, MIN is a term since both the defining block for MIN,

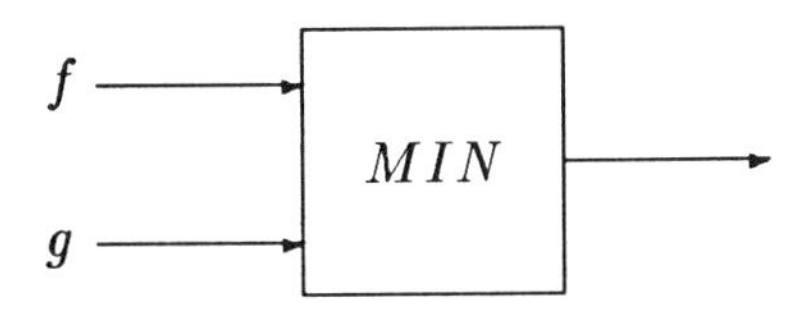

and the following block diagram

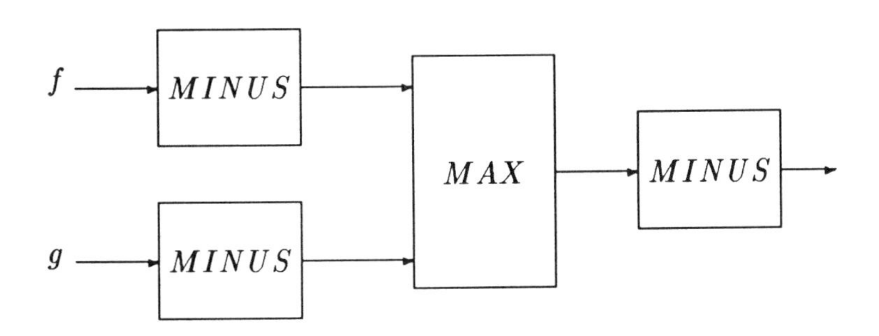

have the same output, $f \wedge g$.

Example 2.5 *Employing f and g as given in Example 2.1, we have*

$$f \wedge g = \begin{pmatrix} 0 & 0 & 0 \\ 1 & -1 & 0 \\ -1 & 0 & 0 \end{pmatrix}_{3,4}^{0}$$

On the other hand, we can calculate $f \wedge g$ *using* $-(-f \vee -g)$. *Accordingly,*

$$-f = \begin{pmatrix} -2 & -3 & 0 \\ -1 & -4 & 0 \\ 1 & -1 & 0 \end{pmatrix}_{3,4}^{0}$$

and

$$-g = \begin{pmatrix} 0 & 0 & 0 \\ -2 & 1 & -3 \\ -1 & 0 & -2 \end{pmatrix}_{3,4}^{0}$$

Consequently,

$$-f \vee -g = \begin{pmatrix} 0 & 0 & 0 \\ -1 & 1 & 0 \\ 1 & 0 & 0 \end{pmatrix}_{3,4}^{0}$$

and finally,

$$-(-f \vee -g) = \begin{pmatrix} 1 & -1 \\ -1 & 0 \end{pmatrix}_{3,3}^{0}$$

Several rectifier-type device operations are also described using the three simple fundamental operations $+$, $\cdot$, and $\vee$. The absolute value function, denoted by ABS or $|\ |$, is a unary operation, that is

$$|\ |: \Re^{Z \times Z} \rightarrow \Re^{Z \times Z}$$

where

$$|\ f\ |(n,m) = |\ f(n,m)\ |$$

Using block diagrams, it too is a term since the defining block

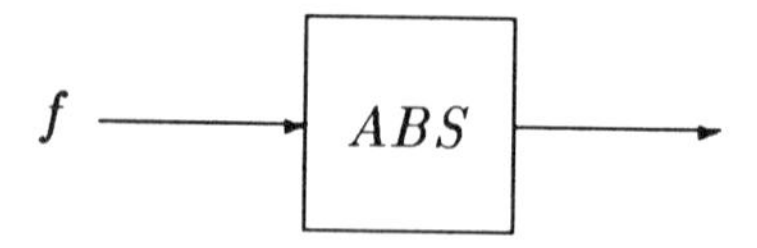

and

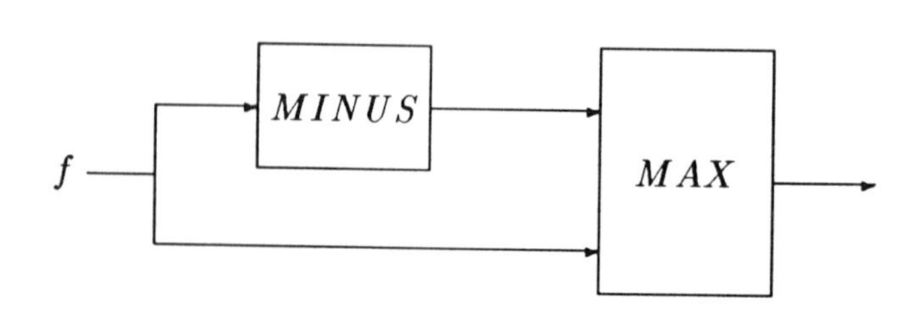

have the same output $| f |$. Equivalently, $| f |= f \vee (-f)$.

Two related unary operators on a digital signal f is the positive part operator and the negative part operator denoted respectively by $(f)^+$ and $(f)^-$. These are defined by

$$(f)^+(n,m) = \begin{cases} f(n,m) & \textit{whenever } f(n,m) > 0 \\ 0 & \textit{otherwise} \end{cases}$$

and

$$(f)^-(n,m) = \begin{cases} -f(n,m) & \textit{whenever } f(n,m) < 0 \\ 0 & \textit{otherwise} \end{cases}$$

Note that the result of applying either the positive part operation or the negative part operation to an arbitrary function results in a function which is never negative in value. The positive part operation has the block diagram

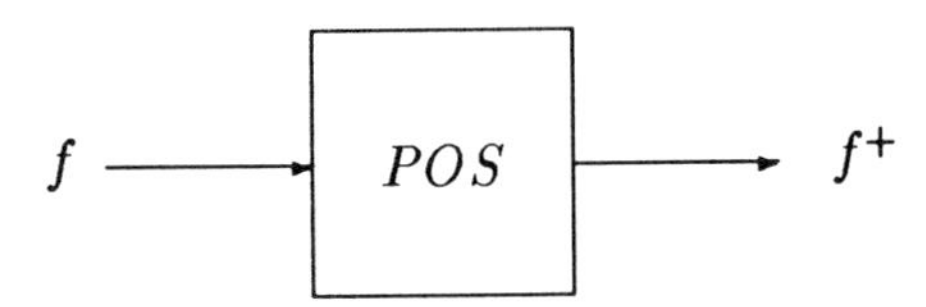

Similarly, the negative part operation has the block representation:

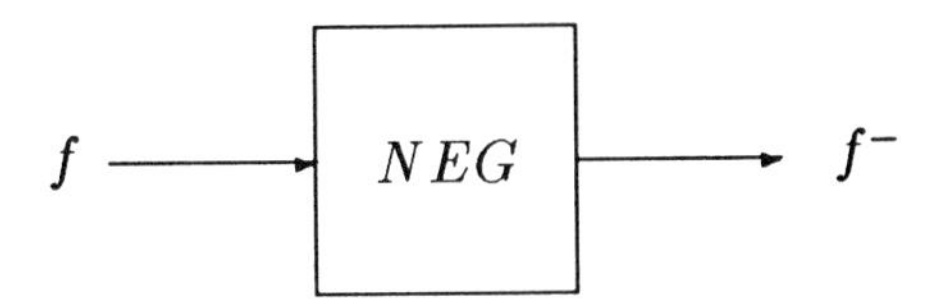

Both of these operations are terms since

$$f^+ = f \vee 0_Z$$

whereas

$$f^- = -(f \wedge 0_Z)$$

Example 2.6 *Again consider*

$$g = \begin{pmatrix} 2 & -1 & 3 \\ 1 & 0 & 2 \end{pmatrix}_{3,4}^{0}$$

then

$$\mid g \mid = \begin{pmatrix} 2 & 1 & 3 \\ 1 & 0 & 2 \end{pmatrix}_{3,4}^{0}$$

and

$$g^{+} = \begin{pmatrix} 2 & 0 & 3 \\ 1 & 0 & 2 \end{pmatrix}_{3,4}^{0}$$

whereas

$$g^{-} = \begin{pmatrix} 0 & 1 & 0 \\ 0 & 0 & 0 \end{pmatrix}_{3,4}^{0}$$

We have shown that the macro operations: SQ, $MINUS$, MIN, ABS, POS and NEG are all terms. This follows since each operation was obtained by using the three range induced fundamental operations along with function composition. It follows that any expression involving the six macro operations above, will equivalently involve the three fundamental range induced operations. In subsequent sections, these macro operations will be used in providing additional terms as well as more advanced algorithms for processing digital signals.

2.3 Additional Important Digital Signals and Macro Operators

The principal purpose of this section is to show how to use the macro operations previously given in order to shape signals. In the shaping process, several operations are often employed with one of their arguments equal to a constant digital signal. Accordingly, it is useful at this time to define two binary macro operations called $SCALAR$ and $OFFSET$. Both of these operations are terms including constant signals as one of their arguments. However, unlike previous terms, both of these operations involve inputs of different types.

Specifically, they both have arguments consisting of a real number and a digital signal. Hence

$$SCALAR : \Re^{Z\times Z} \times \Re \to \Re^{Z\times Z}$$

and

$$OFFSET : \Re^{Z\times Z} \times \Re \to \Re^{Z\times Z}$$

These terms are defined by

$$SCALAR(f,a)(n,m) = a \cdot f(n,m)$$

and

$$OFFSET(f,a)(n,m) = f(n,m) + a$$

Thus, the $SCALAR$ operation is just like scalar multiplication in a vector space. The $OFFSET$ operation "lifts" or offsets the signal by adding the value a to every value of f. The operations, $SCALAR(f,a)$ and $OFFSET(f,a)$ are more commonly denoted by "$a \cdot f$" and "$f + a$", respectively. Both operations are terms since their defining blocks:

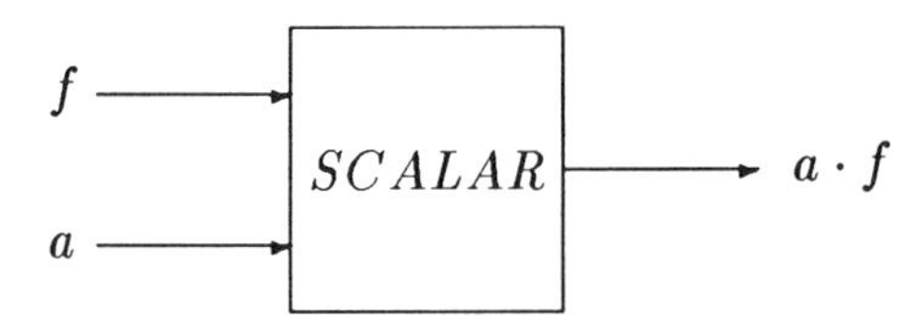

and

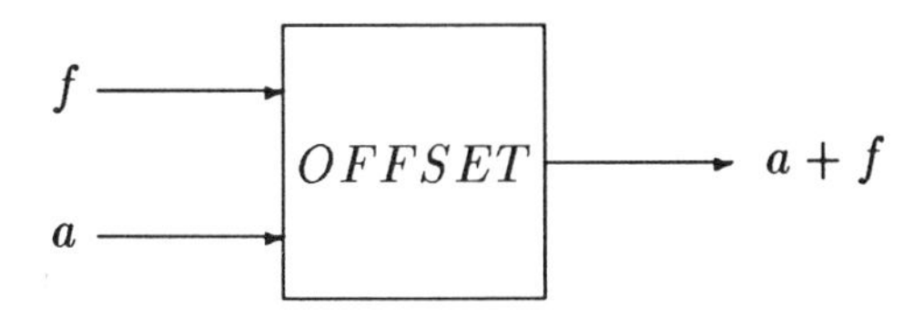

have equal outputs, respectively, with

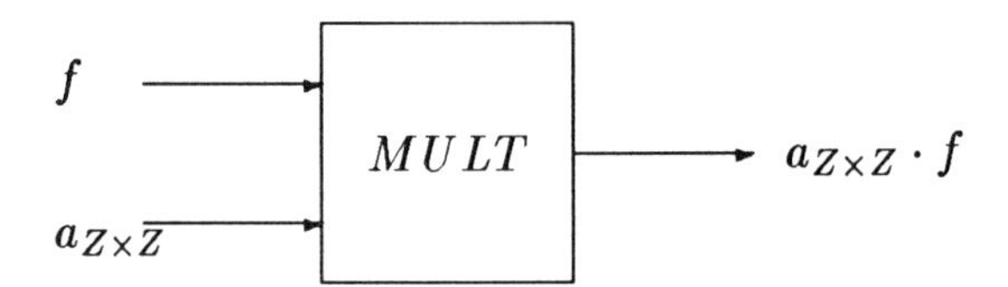

and

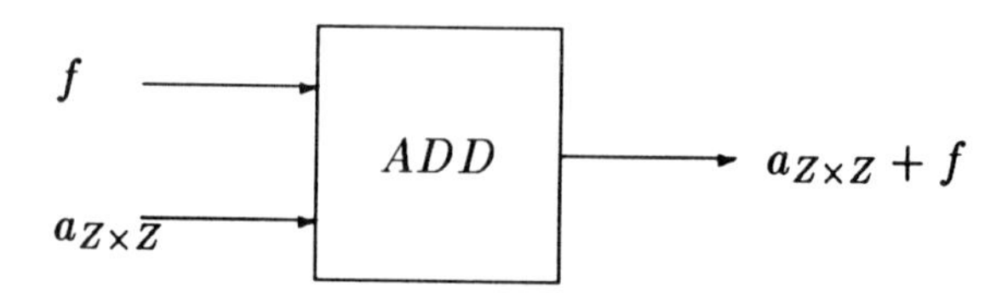

An additional term of great importance is the clipping circuit, denoted by $CLIP$. It has the property that any input signal f will come out of the $CLIP$ operation unaltered for all values less than or equal to some threshold value t. For all values greater than this threshold value, the value t comes out. Thus,

$$CLIP(f,t)(n,m) = \begin{cases} f(n,m) & \textit{wherever } f(n,m) \leq t \\ t & \textit{otherwise} \end{cases}$$

The block diagram illustrative of this operation is given by

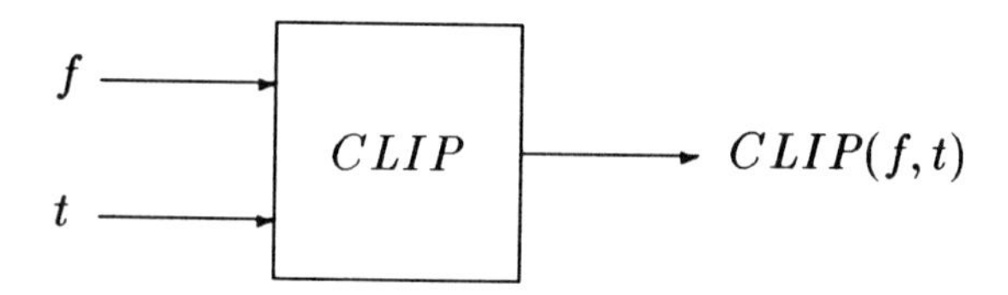

Observe that the output of $CLIP(f,t)$ can be obtained by using the following block diagram:

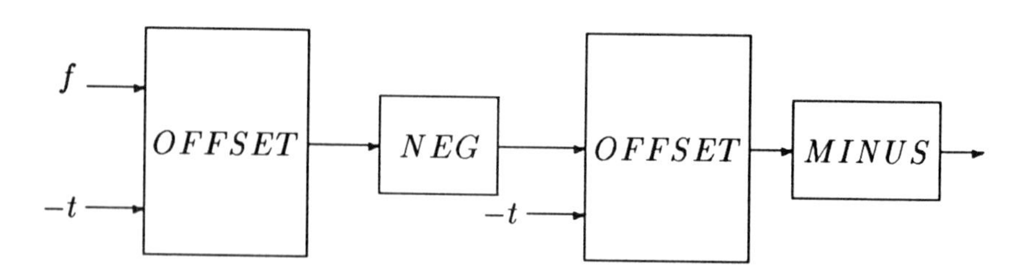

Example 2.7 *Let*

$$f = \begin{pmatrix} 2 & 3 \\ 1 & 4 \\ -1 & 1 \end{pmatrix}_{3,4}^{0}$$

then

$$CLIP(f, 2) = \begin{pmatrix} 2 & 2 \\ 1 & 2 \\ -1 & 1 \end{pmatrix}^{0}_{3,4} = g$$

Now, we will use the last block diagram above to again obtain g. A trace follows.

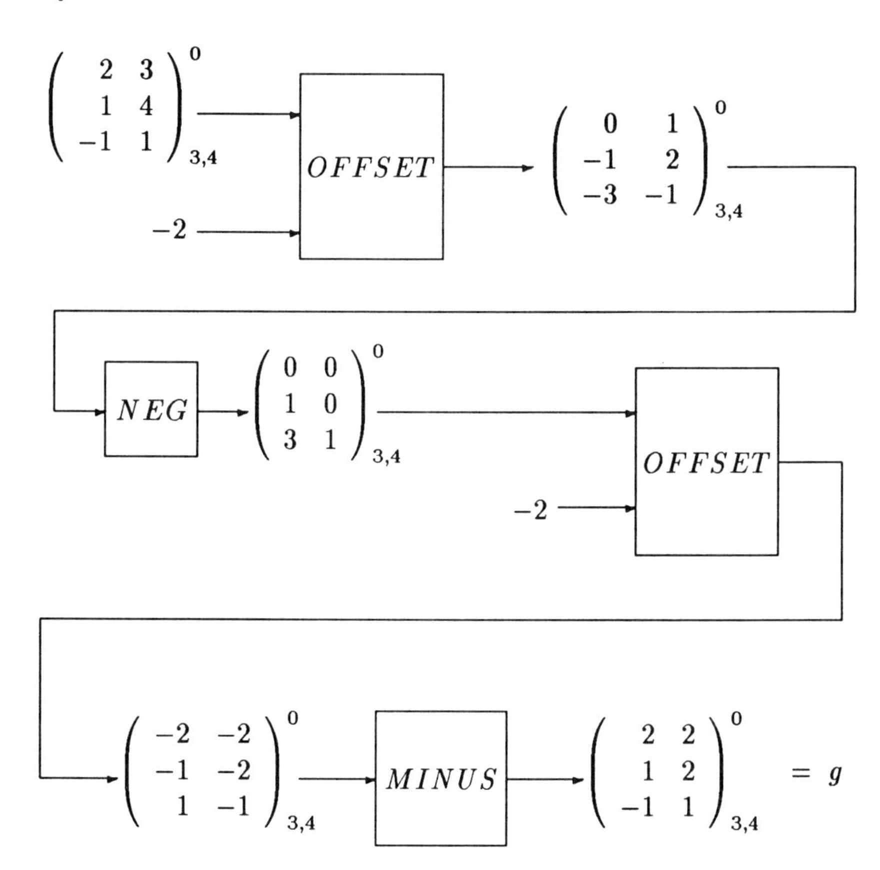

2.4 Equational Identities Involving Range Induced Operations

The very rich algebraic structure in $\Re^{Z \times Z}$ due to the induced operations of pointwise addition, multiplication, and maximum will be further revealed and additionally illustrated. In the following we will let f, g and h be digital signals. Moreover, a and b will denote real values.

The addition operation has the following properties:

G1) Associative Law: $f + (g + h) = (f + g) + h$

G2) Zero Signal Property: $0_{Z\times Z} + f = f + 0_{Z\times Z} = f$

G3) Minus Signal Property: $-f + f = f + (-f) = 0_{Z\times Z}$

G4) Commutative Law: $f + g = g + f$

These properties show that $\Re^{Z\times Z}$ is an Abelian group under addition. Property **G1)** will allow us to use the block diagram for addition as if it is a ternary or higher arty operator. That is, to add three digital signals we will use

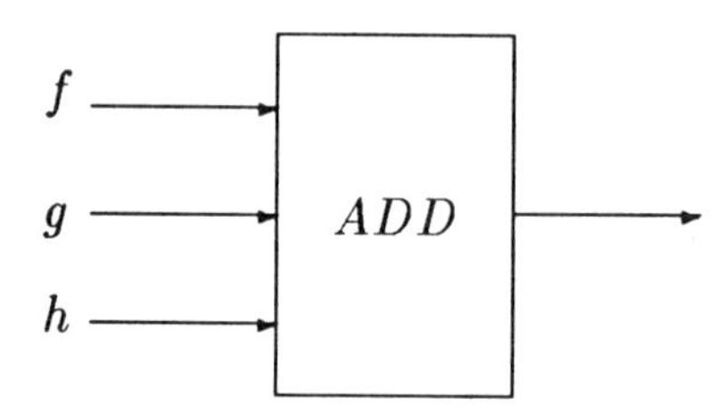

Equivalently, in mathematical terms, we could write $f+g+h$ instead of $f + (g + h)$ or $(f + g) + h$. More generally, for any finite number of digital signals f_1, f_2, ..., f_N, we will write

$$\sum_{i=1}^{N} f_i \quad \text{or} \quad ADD(f_i)$$

instead of $f_1 + f_2 + ... + f_N$. The equivalent block diagram would be

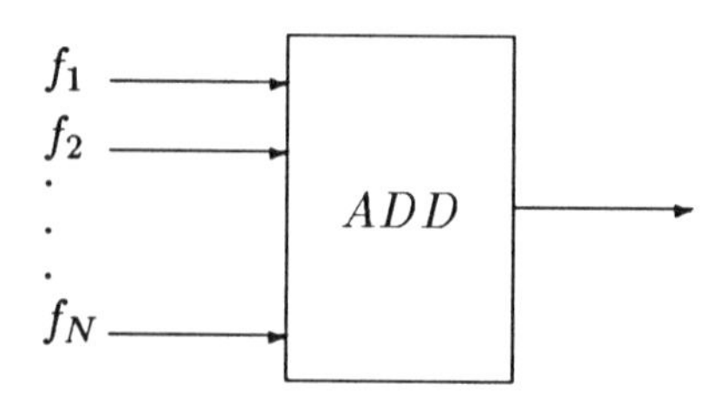

Moreover, the order of summation does not matter due to **G4)**.

Observe also, that the $SCALAR$ operation whose arguments are a real number along with a digital signal satisfies the following:

V1) Associative Law: $(ab)f = a(bf)$
V2) Unity Preservation: $1f = f$
V3) Distributive Law: $a(f + g) = af + ag$
V4) Distributive Law: $(a + b)f = af + bf$

The first distributive law shows that scalar multiplication by reals is distributive with respect to digital signal processing addition. The second distributive law shows that scalar multiplication by digital signals is distributive with respect to the addition of the reals. These two laws can be written in a more informative fashion as

$$SCALAR(ADD(f, g); a) = ADD(SCALAR(f, a), SCALAR(g, a))$$

and

$$SCALAR(f; a + b) = ADD(SCALAR(f, a), SCALAR(f, b))$$

respectively. In any case, **V1)** through **V4)** along with **G1)** through **G4)** shows that $\Re^{Z \times Z}$ is a vector space over the reals.

The multiplication operation in $\Re^{Z \times Z}$ satisfies several important properties:

R1) Associative Law: $(fg)h = f(gh)$
R2) Distributive Laws: $f(g+h) = fg+fh$ and $(f+g)h = fh+gh$
R3) Commutative Law: $fg = gf$
R4) Unity Signal Property: $1_{Z \times Z} \cdot f = f \cdot 1_{Z \times Z} = f$

The multiplicative associative law, being similar to the additive associative law allows us to write, for any number of digital signals $f_1, f_2, \ldots, f_N$

$$\prod_{i=1}^{N} f_i = f_1 \cdot f_2 \cdots \cdot f_N = MULT(f_i)$$

The corresponding block diagram is

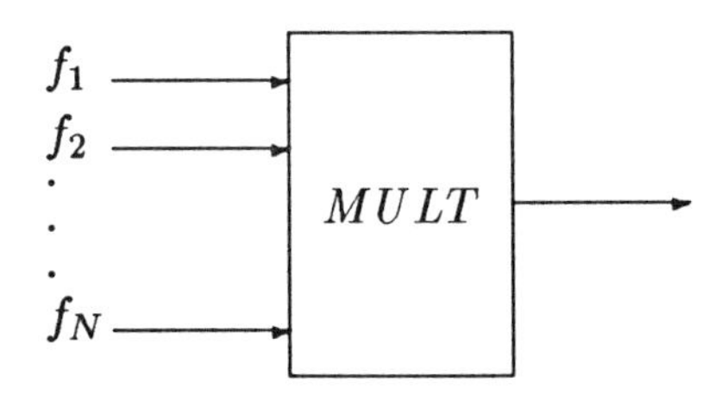

The distributive laws given in **R2)** could be reduced to only one law because of **R3)**.

The system $\Re^{Z\times Z}$ using the above identities along with the group structure is a commutative ring with unity. However, even further structure is present. The *SCALAR* operation commutes with digital signal processing multiplication as the following property **A1)** shows.

A1) Commutative Law: $a(fg) = af \cdot g = f \cdot ag$

Thus, $\Re^{Z\times Z}$ is a commutative, associative algebra with unity. Further algebraic operations in $\Re^{Z\times Z}$ involve the *MAX* and *MIN* operations since the following properties hold:

L1) Commutative Laws:
a.) $f \vee g = g \vee f$
b.) $f \wedge g = g \wedge f$

L2) Associative Laws:
a.) $(f \vee g) \vee h = f \vee (g \vee h)$
b.) $(f \wedge g) \wedge h = f \wedge (g \wedge h)$

L3) Idempotent Laws:
a.) $f \vee f = f$
b.) $f \wedge f = f$

L4) Absorption Laws:
a.) $(f \vee g) \wedge f = f$
b.) $(f \wedge g) \vee f = f$

L5) Distributive Laws:
a.) $f \vee (g \wedge h) = (f \vee g) \wedge (f \vee h)$
b.) $f \wedge (g \vee h) = (f \wedge g) \vee (f \wedge h)$

Again, because of the associative law, when a maximum of a finite number of digital signals $f_1, f_2, ..., f_N$ is desired we will write

$$\bigvee_{i=1}^{N} f_i = f_1 \vee f_2 \vee .... \vee f_N = MAX(f_i)$$

The block diagram illustrating this operation is

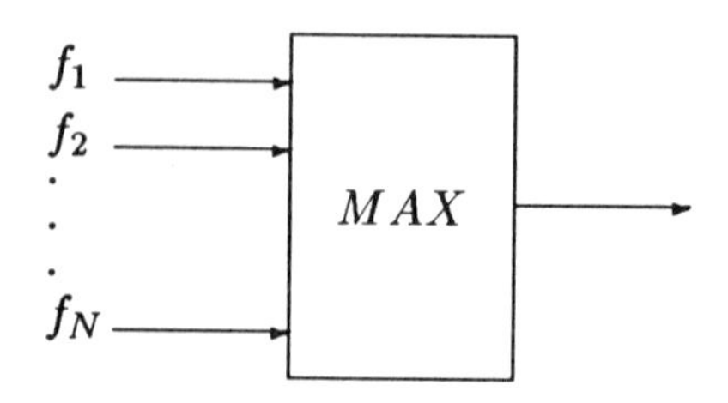

An analogous block diagram can be given for the minimum operation.

The above properties show that $\Re^{Z\times Z}$, along with $\vee$ and $\wedge$, is a distributive lattice. Combining this with the associative algebra structure shows that $\Re^{Z\times Z}$ is a distributive lattice commutative associative algebra with identity. Consequently, the following two properties also hold true:

M1) Max/Min Additive Law: $f+g=(f\vee g)+(f\wedge g)$
M2) Distributive Law: $(f\vee g)+h=(f+h)\vee(g+h)$

Example 2.8 *Here we will illustrate properties* **M1)** *and* **M2)**. *First we let*

$$f = \begin{pmatrix} 2 & 1 & 0 \\ 3 & -1 & 2 \end{pmatrix}_{0,0}^{0}$$

and

$$g = \begin{pmatrix} 4 & -3 & 0 \\ 2 & 1 & 1 \\ 2 & 3 & 0 \end{pmatrix}_{0,1}^{0}$$

then

$$f+g = \begin{pmatrix} 4 & -3 & 0 \\ 4 & 2 & 1 \\ 5 & 2 & 2 \end{pmatrix}_{0,1}^{0}$$

and

$$f\vee g = \begin{pmatrix} 4 & 0 & 0 \\ 2 & 1 & 1 \\ 3 & 3 & 2 \end{pmatrix}_{0,1}^{0}$$

also

$$f\wedge g = \begin{pmatrix} 0 & -3 & 0 \\ 2 & 1 & 0 \\ 2 & -1 & 0 \end{pmatrix}_{0,1}^{0}$$

so

$$(f\vee g)+(f\wedge g) = \begin{pmatrix} 4 & -3 & 0 \\ 4 & 2 & 1 \\ 5 & 2 & 2 \end{pmatrix}_{0,1}^{0}$$

which illustrates property **M1**). *To illustrate property* **M2**) *we first let*

$$h = \begin{pmatrix} 0 & 3 & 4 \\ 2 & 0 & 1 \\ -1 & 2 & 0 \end{pmatrix}_{0,1}^{0}$$

then, using f and g as before we obtain

$$(f \vee g) + h = \begin{pmatrix} 4 & 3 & 4 \\ 4 & 1 & 2 \\ 2 & 5 & 2 \end{pmatrix}_{0,1}^{0}$$

and

$$f + h = \begin{pmatrix} 0 & 3 & 4 \\ 4 & 1 & 1 \\ 2 & 1 & 2 \end{pmatrix}_{0,1}^{0}$$

also

$$g + h = \begin{pmatrix} 4 & 0 & 4 \\ 4 & 1 & 2 \\ 1 & 5 & 0 \end{pmatrix}_{0,1}^{0}$$

so, we obtain

$$(f + h) \vee (g + h) = \begin{pmatrix} 4 & 3 & 4 \\ 4 & 1 & 2 \\ 2 & 5 & 2 \end{pmatrix}_{0,1}^{0}$$

which illustrates property **M2**).

If f and g are in F, that is they have finite support, then

- $f + g \in F$
- $f \vee g \in F$
- $f \cdot g \in F$

Moreover, all the properties just given for functions in $\Re^{Z\times Z}$ hold for functions in F (except **R4**). Accordingly, the structure is a sub-distributive lattice commutative associative algebra. The zero function $0_{Z\times Z}$ in $\Re^{Z\times Z}$ is also in F, but the unity function $1_{Z\times Z}$ in $\Re^{Z\times Z}$ is not in F.

2.5 Fundamental Domain Induced Operations

Digital signals have as their domain the set of lattice points that are pairs of integers. Similar to the way that the algebraic structure of the reals induced operations in $\Re^{Z\times Z}$, group operations in the integers induce operations in $\Re^{Z\times Z}$. Precisely three *domain induced operations* are defined herein. They are the shift, the 90° rotation and the reflection operations.

The first domain induced operation, the shift operation, is denoted by S. This operation exists, and therefore can be applied to any digital signal since every integer has a successor. The operator S is unary, that is

$$S : \Re^{Z\times Z} \rightarrow \Re^{Z\times Z}$$

It is defined by

$$(S(f))(n,m) = f(n-1,m)$$

The shift operation, when applied to a digital signal "moves" it one unit to the right. Accordingly, for any digital signal f, given as the bound matrix

$$f = \begin{pmatrix} & \cdot & \cdot & \cdot & \cdot & \\ & \cdot & \cdot & \cdot & \cdot & \\ & \cdot & \cdot & \cdot & \cdot & \\ \cdots & j & e & d & c & \cdots \\ \cdots & k & f & \boxed{a} & b & \cdots \\ \cdots & l & g & h & i & \cdots \\ & \cdot & \cdot & \cdot & \cdot & \\ & \cdot & \cdot & \cdot & \cdot & \\ & \cdot & \cdot & \cdot & \cdot & \end{pmatrix}$$

we have

$$S(f) = \begin{pmatrix} & \cdot & \cdot & \cdot & \cdot & \\ & \cdot & \cdot & \cdot & \cdot & \\ & \cdot & \cdot & \cdot & \cdot & \\ \cdots & j & e & d & c & \cdots \\ \cdots & k & \boxed{f} & a & b & \cdots \\ \cdots & l & g & h & i & \cdots \\ & \cdot & \cdot & \cdot & \cdot & \\ & \cdot & \cdot & \cdot & \cdot & \\ & \cdot & \cdot & \cdot & \cdot & \end{pmatrix}$$

Finally, if f has finite support, then f can be written as

$$f = \begin{pmatrix} a_{11} & a_{12} & \cdots & a_{1m} \\ a_{21} & a_{22} & \cdots & a_{2m} \\ \cdot & & & \\ \cdot & & & \\ \cdot & & & \\ a_{n1} & a_{n2} & \cdots & a_{nm} \end{pmatrix}_{p,q}^{0}$$

hence,

$$S(f) = \begin{pmatrix} a_{11} & a_{12} & \cdots & a_{1m} \\ a_{21} & a_{22} & \cdots & a_{2m} \\ \cdot & & & \\ \cdot & & & \\ \cdot & & & \\ a_{n1} & a_{n2} & \cdots & a_{nm} \end{pmatrix}_{p+1,q}^{0}$$

Example 2.9 *Let*

$$f = \begin{pmatrix} 2 & 1 \\ 0 & 3 \\ 4 & 1 \end{pmatrix}_{-1,1}^{0} = \begin{pmatrix} 2 & 1 \\ 0 & \boxed{3} \\ 4 & 1 \end{pmatrix}^{0}$$

then

$$S(f) = \begin{pmatrix} 2 & 1 \\ 0 & 3 \\ 4 & 1 \end{pmatrix}_{0,1}^{0} = \begin{pmatrix} 2 & 1 \\ \boxed{0} & 3 \\ 4 & 1 \end{pmatrix}^{0}$$

The functions f and $S(f)$ are illustrated in Figures 2.1(a) and 2.1(b), respectively.

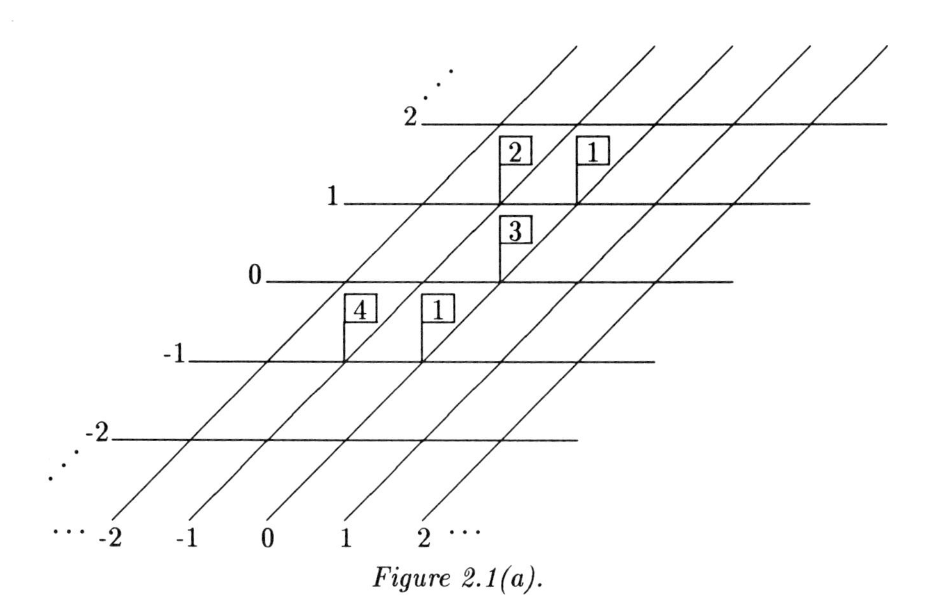

Figure 2.1(a).

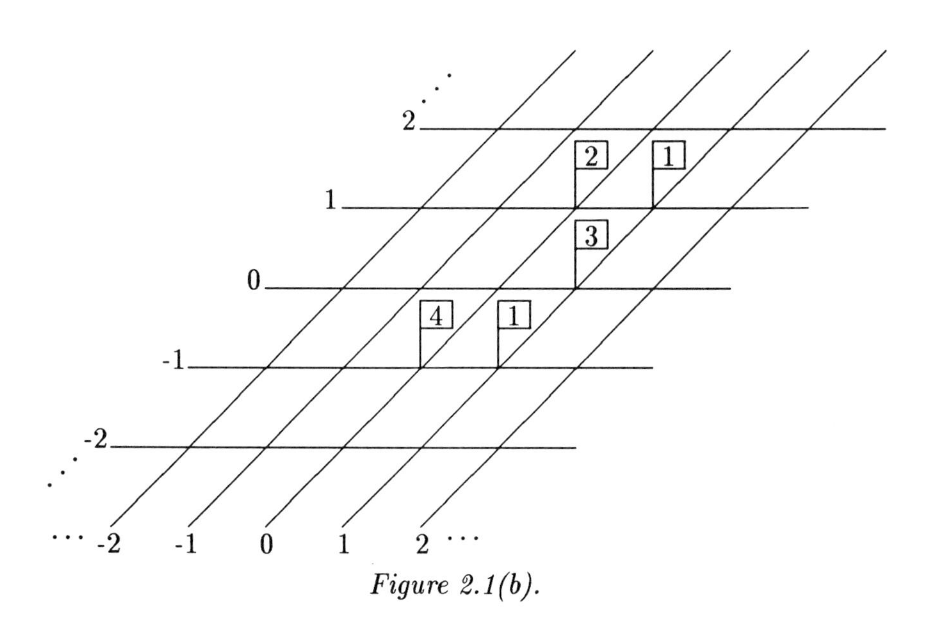

Figure 2.1(b).

Example 2.10 *If we wish to shift the first quadrant step function*

$$u = \begin{pmatrix} \cdot & \cdot & \cdot & \\ \cdot & \cdot & \cdot & \\ \cdot & \cdot & \cdot & \\ 1 & 1 & 1 & \dots \\ 1 & 1 & 1 & \dots \\ \boxed{1} & 1 & 1 & \dots \end{pmatrix}^0$$

then we obtain

$$S(u) = \begin{pmatrix} \cdot & \cdot & \cdot & \\ \cdot & \cdot & \cdot & \\ \cdot & \cdot & \cdot & \\ 0 & 1 & 1 & \dots \\ 0 & 1 & 1 & \dots \\ \boxed{0} & 1 & 1 & \dots \end{pmatrix}^0$$

Example 2.11 *The shift operation performed on the impulse function*

$$\delta = (1)^0_{0,0}$$

results in

$$S(\delta) = (1)^0_{1,0}$$

The shift operation is denoted by the block diagram

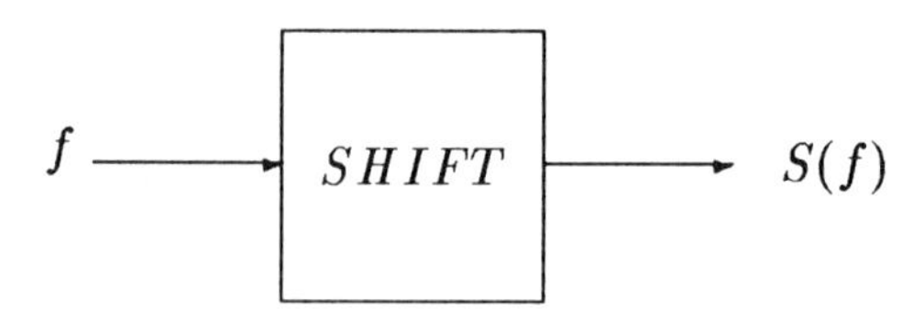

The second domain induced operation, $NINETY$, arises due to the structure of the integral lattice. When the integral lattice is rotated a full 90°, the new lattice which occurs has exactly the same geometric configuration as the original lattice. When applied to a digital signal f, $NINETY(f)$, also symbolized by $N(f)$, has the same values as f each rotated 90° about the origin. Thus,

$$NINETY(f)(i,j) = f(j,-i)$$

Example 2.12 *Say we want to find $N(f)$ where*

$$f = \begin{pmatrix} 2 & 1 \\ 0 & 3 \\ 4 & 1 \end{pmatrix}_{-1,1}^{0}$$

then we can find $N(f)$ "point by point". For instance, using

$$f(-1,-1) = 4$$

we let $j = -1$ and $-i = -1$, this gives

$$N(f)(1,-1) = 4$$

Using

$$f(-1,1) = 2$$

provides the result

$$N(f)(-1,-1) = 2$$

Other values are similarly found, in any case

$$N(f) = \begin{pmatrix} 1 & \boxed{3} & 1 \\ 2 & 0 & 4 \end{pmatrix}^{0}$$

Figures 2.2(a) and 2.2(b) illustrate the functions f and $N(f)$, respectively.

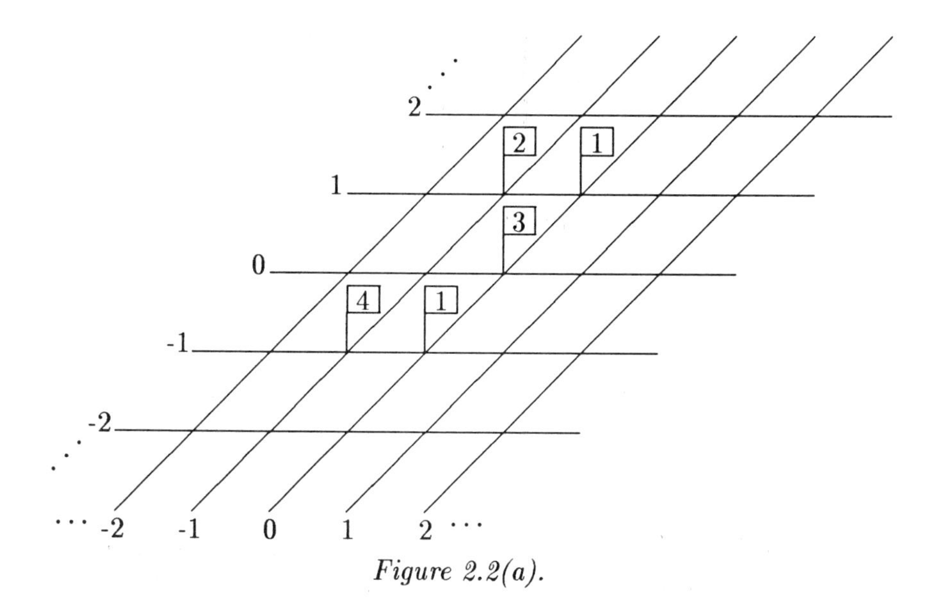

Figure 2.2(a).

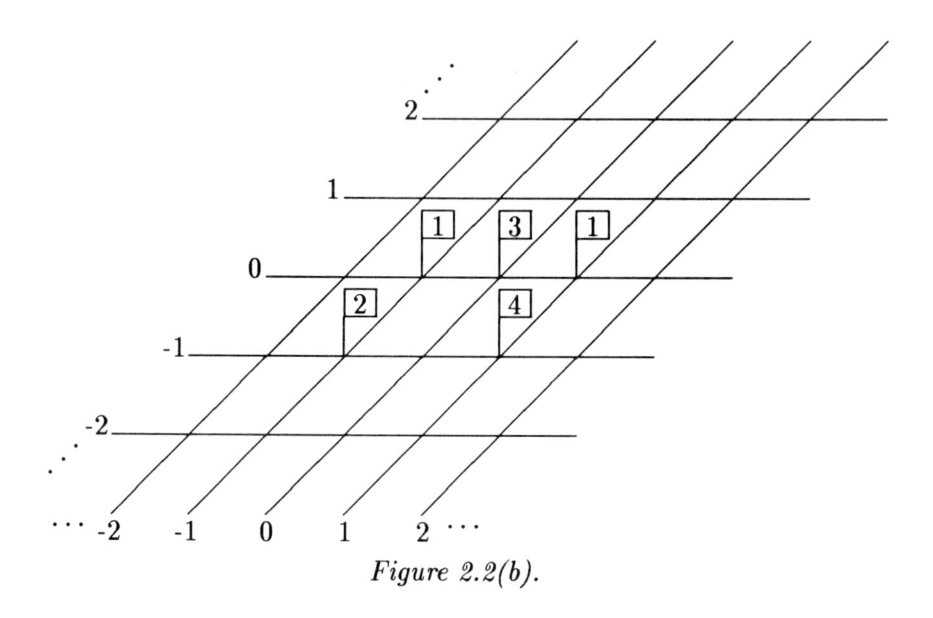

Figure 2.2(b).

Example 2.13 *If the first quadrant unit step function, u, is rotated 90° we obtain*

$$N(u) = \begin{pmatrix} & \cdot & \cdot & \cdot \\ & \cdot & \cdot & \cdot \\ & \cdot & \cdot & \cdot \\ \ldots & 1 & 1 & 1 \\ \ldots & 1 & 1 & \boxed{1} \end{pmatrix}^{0}$$

A block diagram illustration of the 90° rotation is

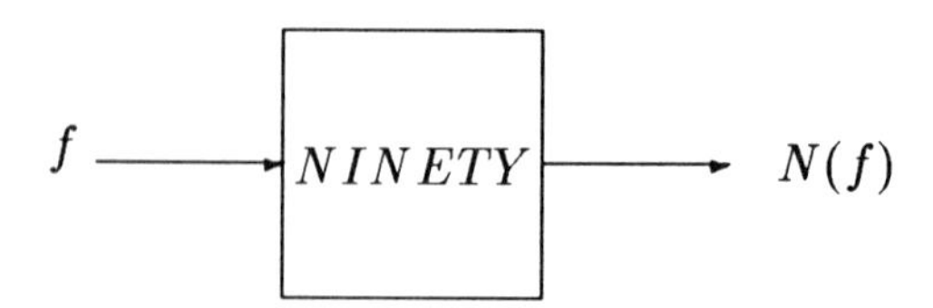

The third and final domain induced operation for bound matrices is a type of matrix transpose, denoted by $REFLECT$, or D. Indeed,

D applied to any digital signal f makes rows of f become columns in $D(f)$ and columns in f become rows in $D(f)$. More precisely, D is defined pointwise using

$$REFLECT(f)(i,j) = f(-j,-i)$$

Thus $D(f)$ can also be obtained by flipping f around a 135° line in the x,y-plane.

Example 2.14 *Let*

$$f = \begin{pmatrix} 2 & 1 \\ 0 & 3 \\ 4 & 1 \end{pmatrix}^{0}_{-1,1}$$

Then using the pointwise definition since

$$f(-j,-i) = f(0,-1) = 1$$

we have

$$D(f)(1,0) = 1$$

Similarly,

$$f(-j,-i) = f(-1,1) = 2$$

so

$$D(f)(-1,1) = 2$$

Accordingly,

$$D(f) = \begin{pmatrix} 2 & 0 & 4 \\ 1 & 3 & 1 \end{pmatrix}^{0}_{-1,1}$$

Observe that $D(f)$ can also be found by placing a mirror along the 135° line and observing f; the resulting digital signal is $D(f)$. The next example will illustrate how it can also be found by rotating f 180° (out of the page) about this line.

Example 2.15 *Let*

$$f = \begin{pmatrix} 2 & 1 \\ -1 & 0 \\ 3 & 0 \end{pmatrix}^{0}_{1,3}$$

$$= \begin{pmatrix} 0 & 0 & 0 & 0 & 2 & 1 \\ 0 & 0 & 0 & 0 & -1 & 0 \\ 0 & 0 & 0 & 0 & 3 & 0 \\ 0 & 0 & 0 & \boxed{0} & 0 & 0 \\ 0 & 0 & 0 & 0 & 0 & 0 \\ 0 & 0 & 0 & 0 & 0 & 0 \end{pmatrix}^{0}$$

Then

$$D(f) = \begin{pmatrix} 0 & 0 & 0 & 0 & 0 & 0 \\ 0 & 0 & 0 & 0 & 0 & 0 \\ 0 & 0 & 0 & 0 & 0 & 0 \\ 0 & 0 & 0 & \boxed{0} & 0 & 0 \\ 2 & -1 & 3 & 0 & 0 & 0 \\ 1 & 0 & 0 & 0 & 0 & 0 \end{pmatrix}$$

$$= \begin{pmatrix} 2 & -1 & 3 \\ 1 & 0 & 0 \end{pmatrix}^{0}_{-3,-1}$$

Example 2.16 *The reflection operation used on the first quadrant unit step function gives*

$$D(u) = \begin{pmatrix} \cdots & 1 & 1 & \boxed{1} \\ \cdots & 1 & 1 & 1 \\ \cdots & 1 & 1 & 1 \\ & \cdot & \cdot & \cdot \\ & \cdot & \cdot & \cdot \\ & \cdot & \cdot & \cdot \end{pmatrix}^{0}$$

A block diagram illustrating the (diagonal) reflection operation is

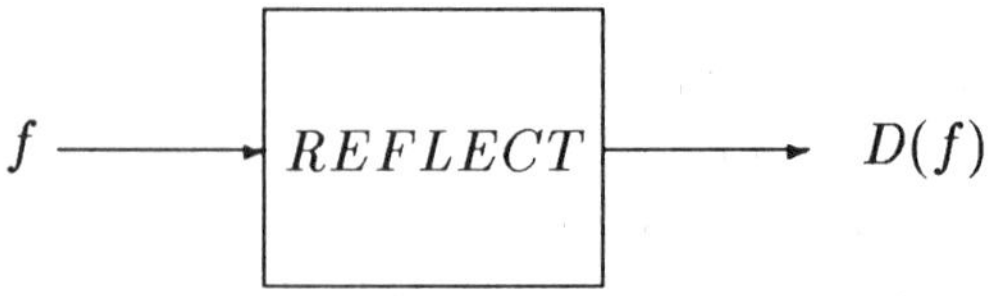

f → *REFLECT* → $D(f)$

2.6 Terms Involving Fundamental Domain Induced Operations

Successive applications of the shift operation produces a translation (to the right) type operation. Indeed, if we let S^n denote n successive function compositions of S, then

$$(S^n(f))(k,j) = f(k-n,j)$$

Note that $n \geq 0$ and $S^0(f) = f$. This amounts to translating the bound vector n places to the right. In terms of block diagrams, we have

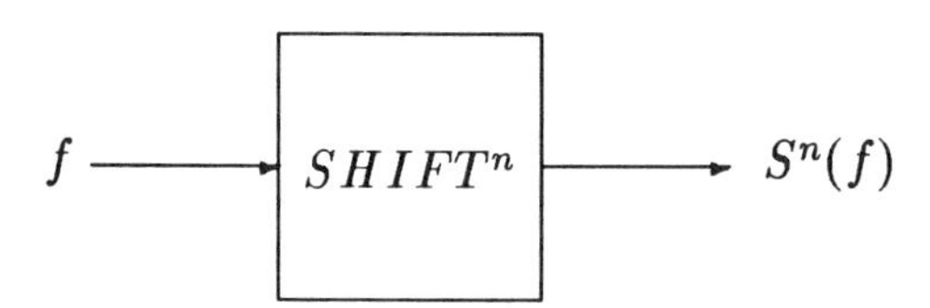

and

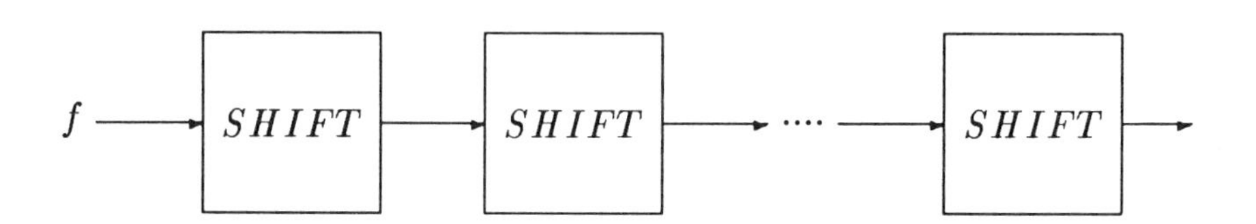

both possessing identical outputs.

Example 2.17 *Suppose that the first quadrant unit step function u is given*

$$u = \begin{pmatrix} \cdot & \cdot & \\ \cdot & \cdot & \\ \cdot & \cdot & \\ 1 & 1 & \ldots \\ \boxed{1} & 1 & \ldots \end{pmatrix}^0$$

then

$$S^3(u) = \begin{pmatrix} \cdot & \cdot & \cdot & \cdot & \cdot & \\ \cdot & \cdot & \cdot & \cdot & \cdot & \\ \cdot & \cdot & \cdot & \cdot & \cdot & \\ 0 & 0 & 0 & 1 & 1 & \ldots \\ \boxed{0} & 0 & 0 & 1 & 1 & \ldots \end{pmatrix}^0$$

Successive applications of the ninety degree rotation produces various other rotations. The 180° rotation of f is given by $N^2(f)$; the 270° rotation of f is given by $N^3(f)$. These rotations can also be obtained pointwise using

$$N^2(f)(i,j) = f(-i,-j)$$

and

$$N^3(f)(i,j) = f(-j,i)$$

Example 2.18 *Let*

$$f = \begin{pmatrix} 2 & -1 \end{pmatrix}^{0}_{3,4}$$

then

$$N^2(f) = \begin{pmatrix} -1 & 2 \end{pmatrix}^{0}_{-4,-4}$$

and

$$N^3(f) = \begin{pmatrix} 2 \\ -1 \end{pmatrix}^{0}_{4,-3}$$

The last result follows since

$$f(-j,i) = f(3,4) = 2$$

hence

$$N^3(f)(4,-3) = 2$$

Similarly

$$f(-j,i) = f(4,4) = -1$$

hence

$$N^3(f)(4,-4) = -1$$

Block diagrams illustrating these operations are

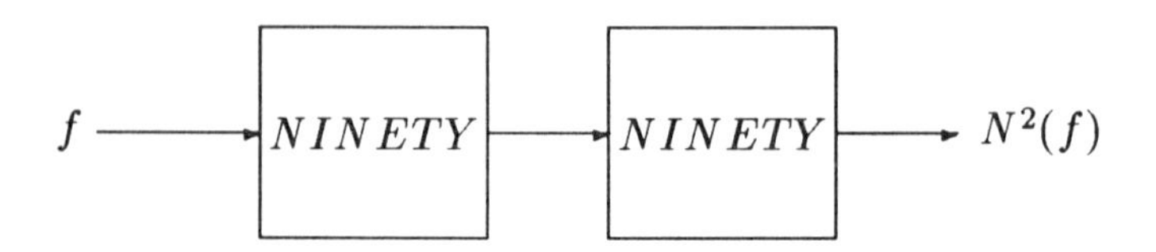

or

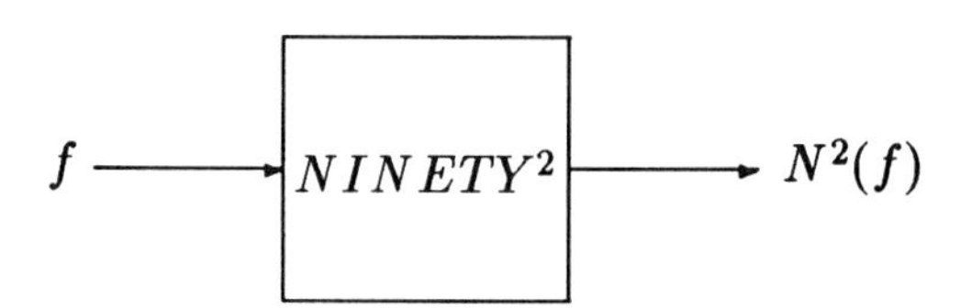

and

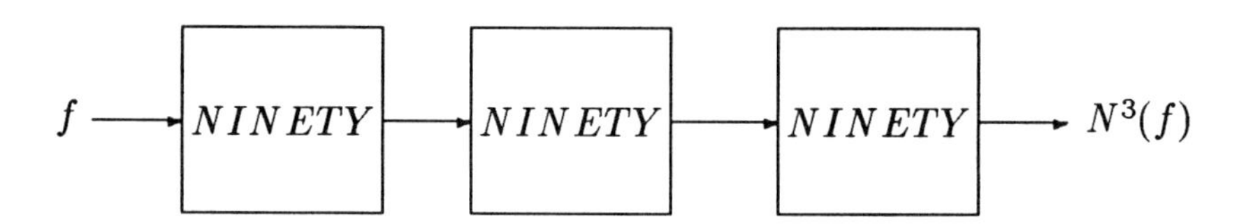

or

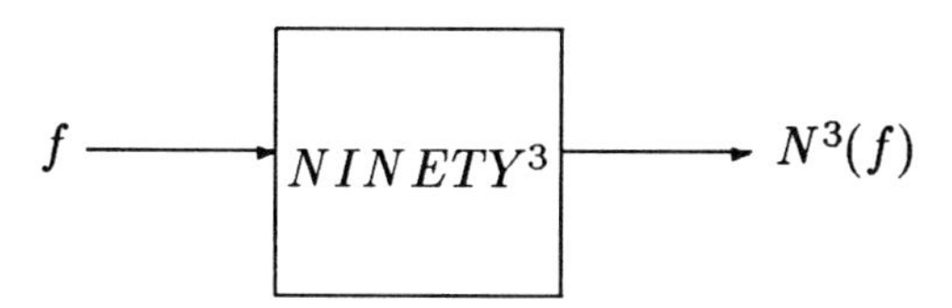

What is of additional consequence is that a general translation operation can be formed using the $SHIFT$ and $NINETY$ operations. Indeed, if we define the translation operation, denoted by $TRAN$ or T, where

$$T : \Re^{Z \times Z} \times \mathbf{Z} \times \mathbf{Z} \rightarrow \Re^{Z \times Z}$$

$$T(f; n, m)(i, j) = f(i - n, j - m)$$

then we also have

$$T(f; n, m) = \begin{cases} NS^m N^3 S^n(f) & n \geq 0, m \geq 0 \\ NS^m NS^{-n} N^2(f) & n < 0, m \geq 0 \\ N^3 S^{-m} NS^n(f) & n \geq 0, m < 0 \\ N^3 S^{-m} N^3 S^{-n} N^2(f) & n < 0, m < 0 \end{cases}$$

Thus, $TRAN$ is a term. It is represented using the following block diagram.

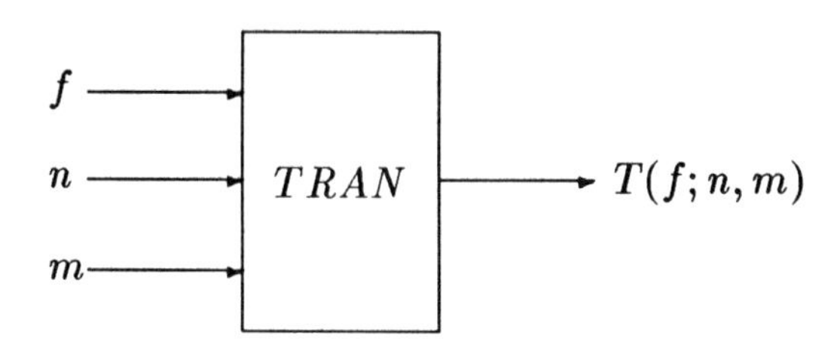

Example 2.19 *Suppose that*

$$f = \begin{pmatrix} 3 \\ 2 \end{pmatrix}^{0}_{1,1}$$

then we have

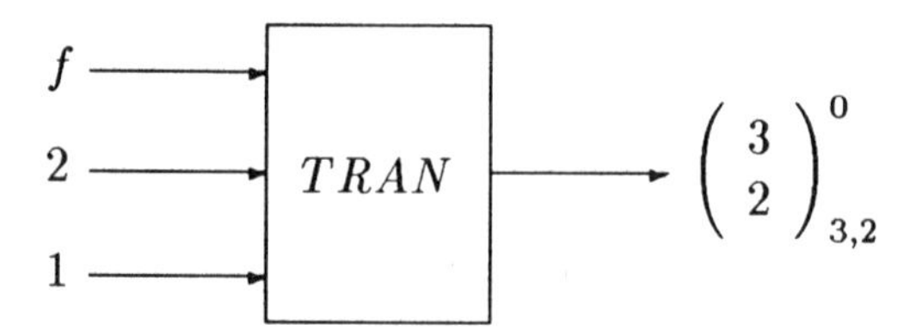

The same result can be found using

$$T(f;2,1) = NSN^3S^2(f)$$

Indeed,

$$S^2(f) = \begin{pmatrix} 3 \\ 2 \end{pmatrix}_{3,1}$$

and

$$N^3S^2(f) = \begin{pmatrix} 2 & 3 \end{pmatrix}_{0,-3}$$

$$SN^3S^2(f) = \begin{pmatrix} 2 & 3 \end{pmatrix}_{1,-3}$$

$$NSN^3S^2(f) = \begin{pmatrix} 3 \\ 2 \end{pmatrix}_{3,2}$$

The translation operation is so important that additional notation shall be employed. Instead of writing $T(f; n, m)$, we will write $f_{n,m}$ for the translate of f by n units to the right and m units up. Thus, the first quadrant unit step function u, and the impulse function δ, when translated n units to the right and m units up are given respectively by $u_{n,m}$ and $\delta_{n,m}$. Thus, we have

$$u_{n,m}(i,j) = u(i-n, j-m)$$

and

$$\delta_{n,m}(i,j) = u(i-n, j-m)$$

Translations of the digital signals u and δ are given in the following example.

Example 2.20

$$u_{-2,-1} = \begin{pmatrix} \cdot & \cdot & \cdot & \cdot & \\ \cdot & \cdot & \cdot & \cdot & \\ \cdot & \cdot & \cdot & \cdot & \\ 1 & 1 & 1 & 1 & \cdots \\ 1 & 1 & \boxed{1} & 1 & \cdots \\ 1 & 1 & 1 & 1 & \cdots \end{pmatrix}^{0}$$

and

$$\delta_{-1,-2} = \begin{pmatrix} & \cdot & \cdot & \cdot & \cdot & \\ & \cdot & \cdot & \cdot & \cdot & \\ & \cdot & \cdot & \cdot & \cdot & \\ \cdots & 0 & 0 & 0 & 0 & \cdots \\ \cdots & 0 & 0 & \boxed{0} & 0 & \cdots \\ \cdots & 0 & 0 & 0 & 0 & \cdots \\ \cdots & 0 & 1 & 0 & 0 & \cdots \\ \cdots & 0 & 0 & 0 & 0 & \cdots \\ & \cdot & \cdot & \cdot & \cdot & \\ & \cdot & \cdot & \cdot & \cdot & \\ & \cdot & \cdot & \cdot & \cdot & \end{pmatrix}^{0}$$

Various other reflection type operations exist besides the transpose type reflection D. Specifically, there is a horizontal, a vertical, and a 45° diagonal reflection. These are respectively denoted by HOR, $VERT$, and $DIFLIP$. They are defined by:

$$HOR(f)(i,j) = f(i,-j)$$

$$VERT(f)(i,j) = f(-i,j)$$

and

$$DIFLIP(f)(i,j) = f(j,i)$$

The following block diagrams illustrate these operations and also show that they are terms. The output of the following two block diagrams is $HOR(f)$.

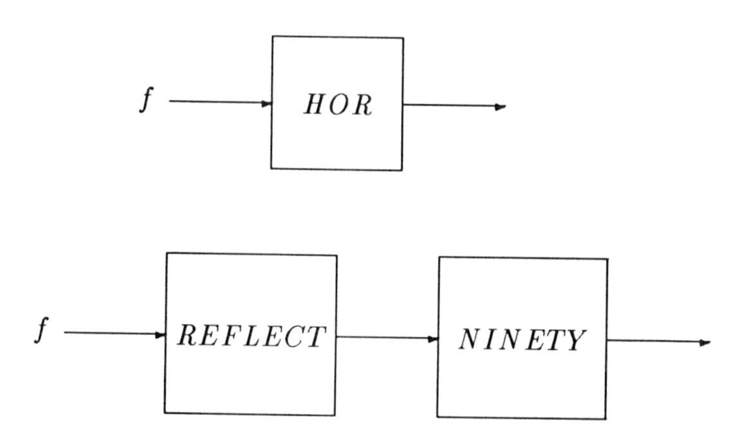

The output of each of the following two block diagrams is $VERT(f)$.

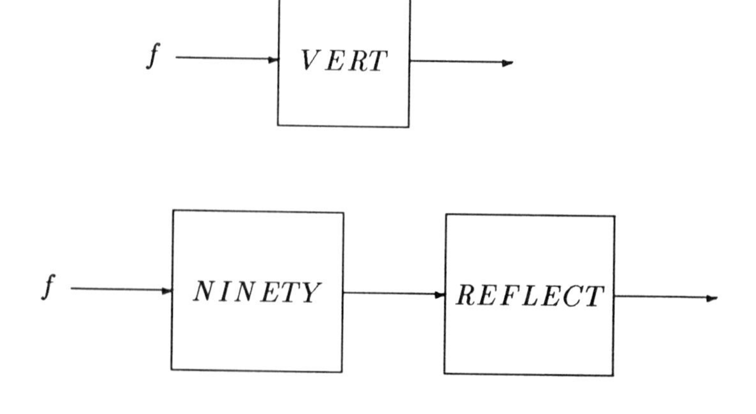

The output of each of the following two block diagrams is $DIFLIP(f)$.

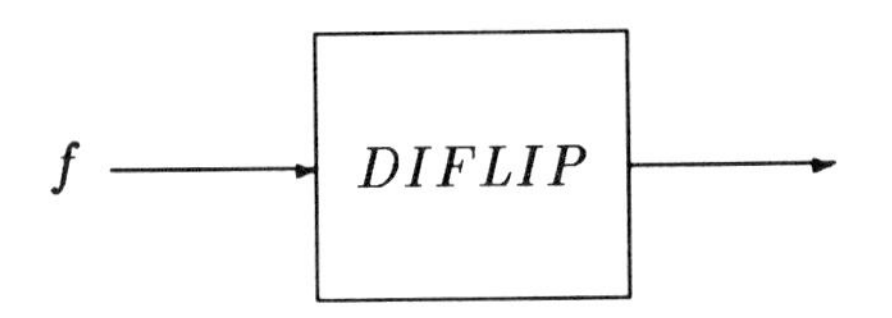

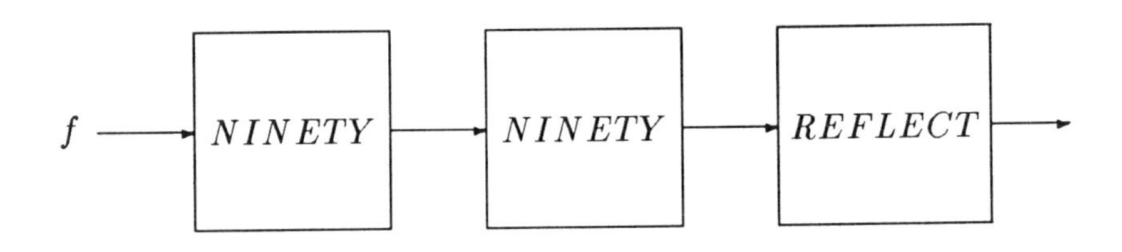

Example 2.21 *If we again let*

$$f = \begin{pmatrix} 3 \\ 2 \end{pmatrix}_{1,1}^{0}$$

as in Example 2.19, *then*

$$D(f) = \begin{pmatrix} 3 & 2 \end{pmatrix}_{-1,-1}^{0}$$

and

$$HOR(f) = N\,D(f) = \begin{pmatrix} 2 \\ 3 \end{pmatrix}_{1,0}^{0}$$

Also,

$$N(f) = \begin{pmatrix} 3 & 2 \end{pmatrix}_{-1,1}^{0}$$

and

$$VERT(f) = D\,N(f) = \begin{pmatrix} 3 \\ 2 \end{pmatrix}_{-1,1}^{0}$$

Finally,

$$N^2(f) = \begin{pmatrix} 2 \\ 3 \end{pmatrix}_{-1,0}^{0}$$

and

$$DIFLIP = D\,N^2(f) = \begin{pmatrix} 2 & 3 \end{pmatrix}^{0}_{0,1}$$

The only possible configuration which was not found using combinations of operations N and D on f, thus far, is

$$N^3(f) = \begin{pmatrix} 2 & 3 \end{pmatrix}^{0}_{0,-1}$$

In the previous example, an image f was used in illustrating HOR, $VERT$, $DIFLIP$ reflection type operations and the N^3 operations. In doing this, we performed all possible combinations of ninety degree rotations and reflections.

2.7 Further Equational Identities

In section 2.4, it was seen that the structure $\Re^{Z\times Z}$ with operations $\cdot, +$, and $\vee$ is a distributed lattice, commutative associative algebra with identity. Now, we will present several properties involving the domain induced operations N, S, D, and also T along with the range induced operations.

To begin, notice that a double application of D to any digital signal always results in the original signal. That is, property

I1) Involution of D:

$$D\,D(f) = f$$

holds true. Accordingly, D is said to be involutory. The same is true for N^2, thus we have

I2) Involution of N^2:

$$N^2\,N^2(f) = N^4(f) = f$$

The translation operation also satisfies the obvious property

C1) Commutative Law:

$$T(T(f;n,m),i,j) = T(T(f;i,j),n,m) = T(f;n+i,m+j)$$

Of most importance is the theorem that domain induced and range induced operations commute. This fact, involving mixed operations, is proved in Reference [1]. In any case, the following properties result from this theorem.

P1) $N(f+g) = N(f) + N(g)$
P2) $N(f \cdot g) = N(f) \cdot N(g)$
P3) $N(f \vee g) = N(f) \vee N(g)$
P4) $N(f \wedge g) = N(f) \wedge N(g)$
P5) $D(f+g) = D(f) + D(g)$
P6) $D(f \cdot g) = D(f) \cdot D(g)$
P7) $D(f \vee g) = D(f) \vee D(g)$
P8) $D(f \wedge g) = D(f) \wedge D(g)$
P9) $T(f+g; n, m) = T(f; n, m) + T(g; n, m)$
P10) $T(f \cdot g; n, m) = T(f; n, m) \cdot T(g; n, m)$
P11) $T(f \vee g; n, m) = T(f; n, m) \vee T(g; n, m)$
P12) $T(f \wedge g; n, m) = T(f; n, m) \wedge T(g; n, m)$

The block diagrams illustrating properties **P1)** through **P4)** are very similar, and the same is true for the block diagram corresponding to property **P5)** through **P8)**, and **P9)** through **P12)**. Accordingly, a representative block diagram shall be chosen from each of these.

Property **P2)** holds true since the following two block diagrams have the same output.

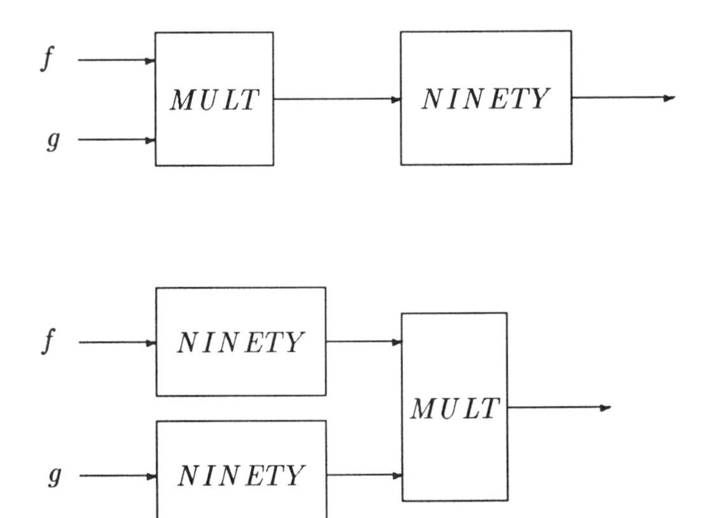

Property **P8)** holds true since the following two block diagrams always have identical outputs.

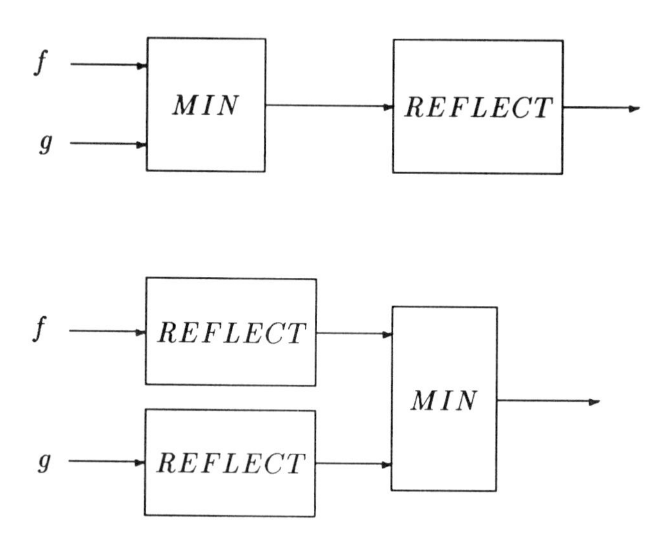

Property **P11)** holds true since the following two block diagrams have the same output.

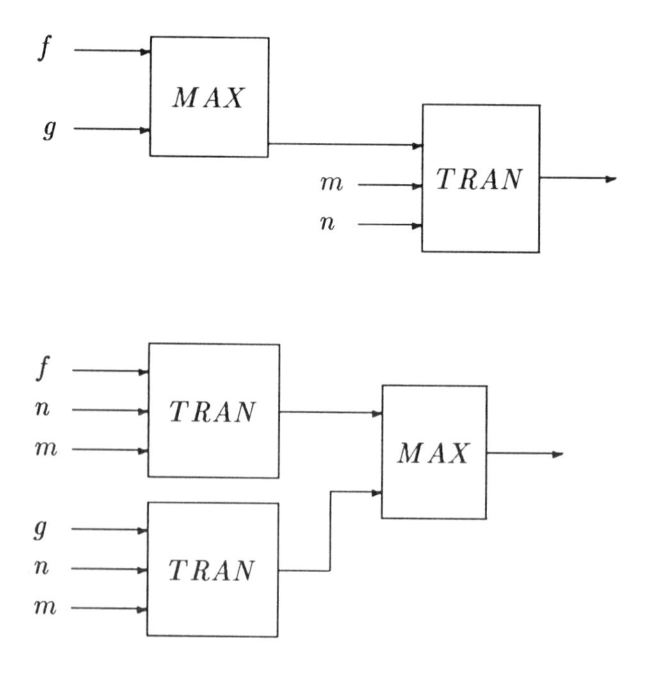

A further consequence of the commutativity between range and domain induced operations is that range induced terms commute with domain induced terms. For instance,

$$HOR \mid f \mid = \mid HOR(f) \mid$$

and

$$TRAN(CLIP(f;t);n,m) = CLIP(TRAN(f;n,m),t)$$

2.8 Exercises

1. Suppose that the two digital signals

$$f = \begin{pmatrix} 3 & 2 & 1 \\ 4 & 1 & 0 \end{pmatrix}_{0,0}^{0}$$

and

$$g = \begin{pmatrix} 1 & 2 \\ 0 & -1 \\ 3 & 2 \end{pmatrix}_{0,0}^{0}$$

a) Write f and g both as 3 by 3 bound matrices, starting at $(0,0)$.

b) Show that

$$f + g = \begin{pmatrix} 4 & 4 & 1 \\ 4 & 0 & 0 \\ 3 & 2 & 0 \end{pmatrix}^{0}$$

c) Find $f \cdot g$.

d) Show that

$$f \vee g = \begin{pmatrix} 3 & 2 & 1 \\ 4 & 1 & 0 \\ 3 & 2 & 0 \end{pmatrix}_{0,0}^{0}$$

2. Let the digital signals f and g be given as

$$f = \begin{pmatrix} 3 & 2 & 1 \end{pmatrix}_{0,0}^{2}$$

$$g = \begin{pmatrix} 4 \\ \boxed{-1} \\ 2 \end{pmatrix}^{-3}$$

Find:

a) $f + g$
b) $f \cdot g$
c) $f \vee g$
d) $f \wedge g$

3. Consider the following block diagram:

$$\begin{pmatrix} 3 & 2 \\ 1 & 4 \\ 2 & 1 \end{pmatrix}^{0}_{0,1}$$

$$\begin{pmatrix} 3 & 1 & 4 \\ 1 & 2 & -1 \end{pmatrix}_{0,0}$$

OPERATION

Output Signal

a) If $OPERATION = ADD$ find the output signal.
b) Show that, if $OPERATION = MULT$ then the output signal is

$$\begin{pmatrix} 0 & 0 & 0 \\ \boxed{3} & 4 & 0 \\ 2 & 2 & 0 \end{pmatrix}^{0}$$

c) If $OPERATION = MAX$ what is the output signal?

4. For each of the signals $f = 3_{Z \times Z}$, δ, and u, defined in Section 1.2, find and illustrate the output of each of the following block diagrams:

a)

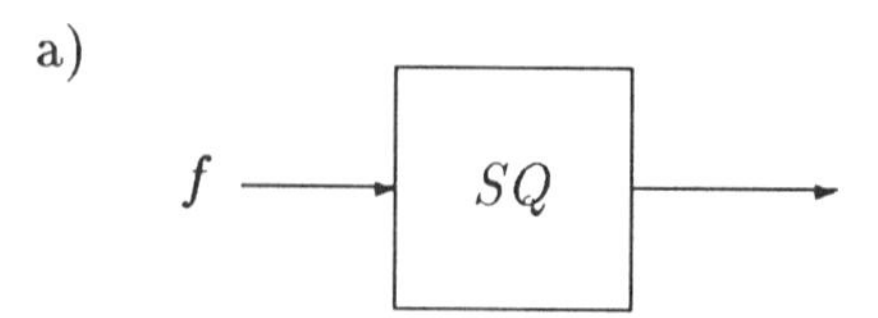

b)

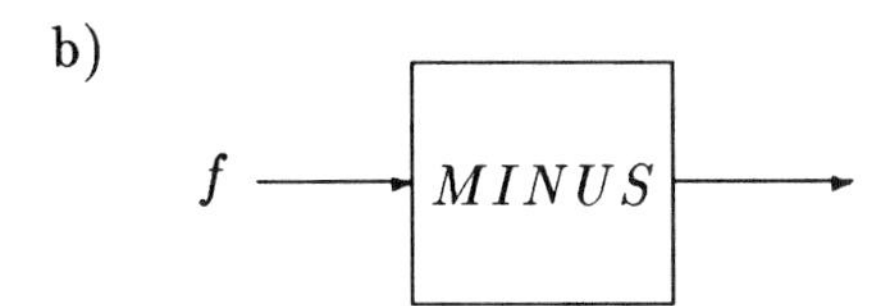

c)

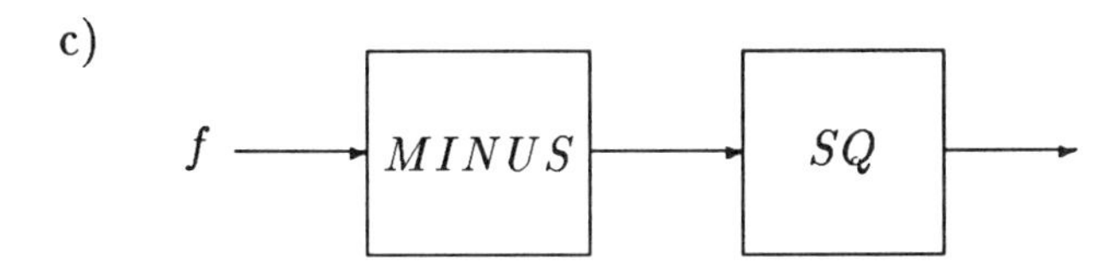

d)

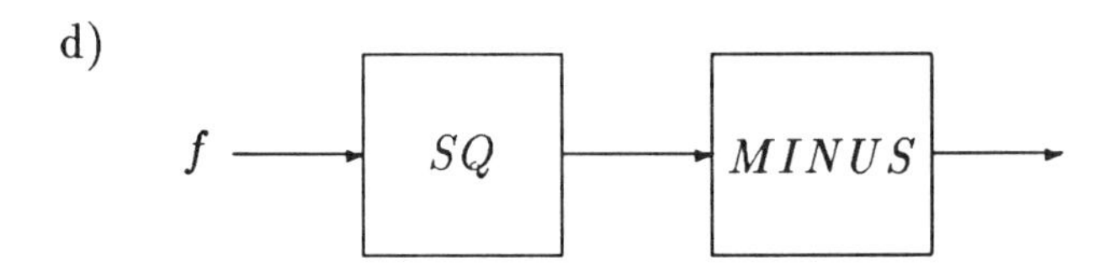

5. Give an example of a digital signal f for which $f \cdot u = u$. Can f have finite support? Explain.

6. A full wave rectifier is defined as ABS, whereas a half wave rectifier is defined as POS. What class of digital signals, when full wave rectified, and half wave rectified yield a common output?

7. Are there any digital signals f which yield the same output to the following block diagrams? If so, find them.

a)

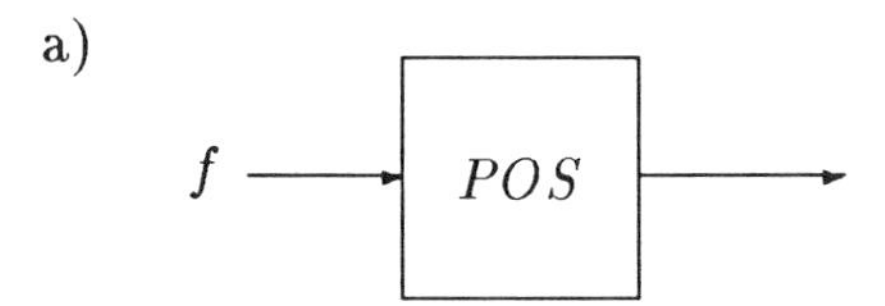

b)

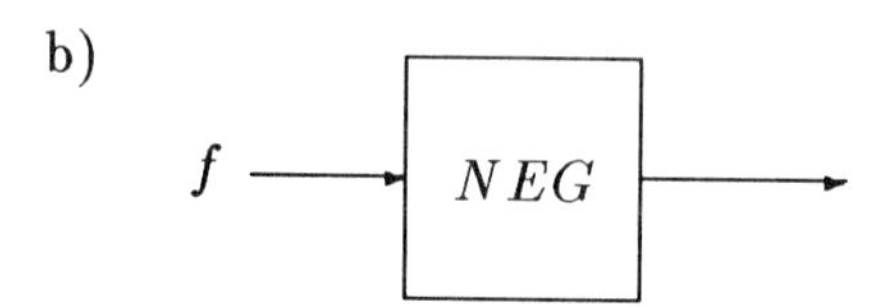

8. Let

$$f = \begin{pmatrix} 3 & 2 \\ -1 & 2 \\ 4 & 3 \end{pmatrix}_{3,4}^{0}$$

Find $CLIP(f;2)$ Do this using the pointwise formulation as well as using the parallel block diagram method.

9. Suppose that the operation $BCLIP$ is given by

$$BCLIP(f;t)(i,j) = \begin{cases} f(i,j) & f(i,j) \geq t \\ t & f(i,j) < t \end{cases}$$

Using f of Example 8, find $BCLIP(f;-1)$. Provide a parallel block diagram formulation of $BCLIP$.

10. The distributive law **V3)** $a(f+g) = af + ag$, holding in the vector space $\Re^{Z \times Z}$, can be described in terms of block diagrams. Indeed, the output of the following two block diagrams are equal.

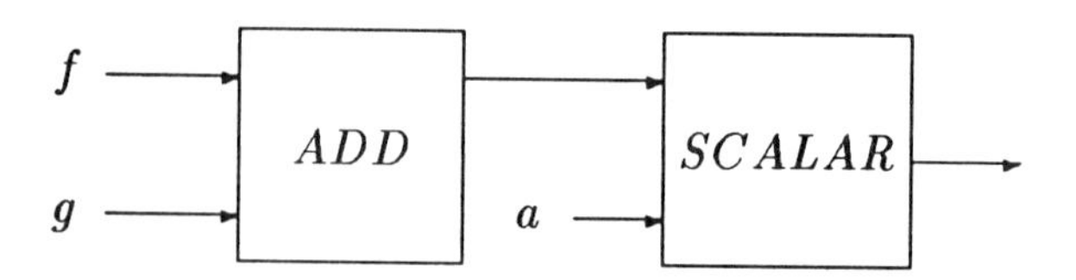

and

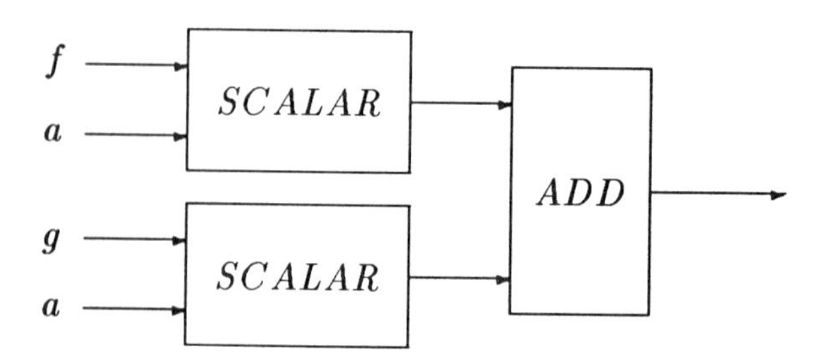

Provide a block diagram illustrating the distributive law **V4)**. Note that in **V4)** a and b are real numbers and therefore $a + b$ is the addition of two real numbers.

11. Give a block diagram illustrating the property **M1)** $f + g = (f \vee g) + (f \wedge g)$ which holds true in $\Re^{Z\times Z}$ since it is a distributive lattice commutative associative algebra with identity.

12. Which of the following are always true for digital signals f ?

    a) $|f| = f^+ + f^-$
    b) $|f| = (f^+ \vee f^-) + (f^+ \wedge f^-)$
    c) $|f| = f^+ \vee f^-$

13. Notice that a horizontally even digital signal is any function f in $\Re^{Z\times Z}$ such that

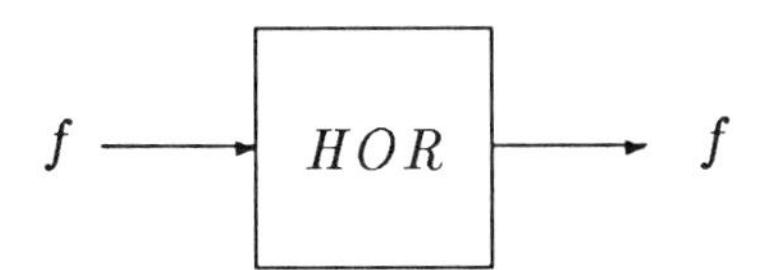

That is, f equals $HOR(f)$, or from the more usual pointwise formula $f(n,m) = f(n,-m)$. Give an example of several such signals, also give a block diagram illustrating a horizontal odd digital signal g defined as $g(n,m) = -g(n,-m)$.

14. For any digital signal f, let $g = \frac{1}{2}(f + HOR(f))$, show that g is an even horizontal signal. Also, if

$$h = \frac{1}{2}(f - HOR(f))$$

show that h is horizontally odd; g is called the horizontal even part of f and h is called the horizontal odd part of f .

15. Define a vertically even digital signal and a vertically odd digital signal similar to what was given for horizontal signals in Exercise 2.13.

16. Define vertical even and odd parts of f.

17. For which types of digital signals f does the following hold true?

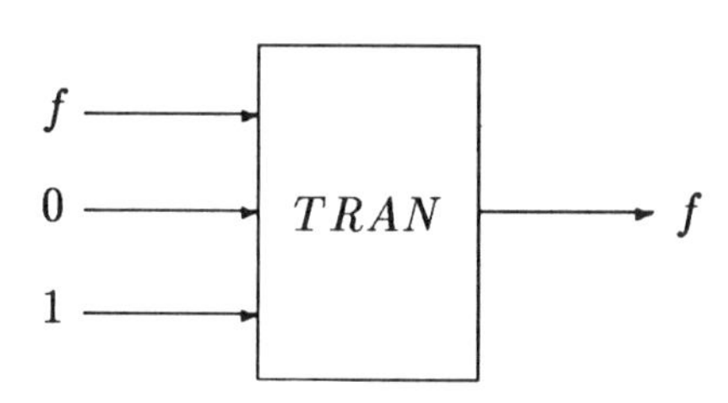

18. Let

$$f = \begin{pmatrix} 2 & 1 \\ 3 & 1 \\ 4 & 2 \end{pmatrix}_{0,0}^{0}$$

and

$$g = \begin{pmatrix} 2 & 1 & 3 \\ 3 & 1 & 2 \end{pmatrix}_{0,0}^{0}$$

Illustrate, using f and g, the commutativity of the domain and range induced operation **P1)** through **P12)** given in section 2.7. For **P9)** through **P12)** use $n = 2$, $m = 1$.

3

Convolution of Digital Signals

3.1 The Support Region for Convolution

In this section, we will assume that all digital signals f are of finite support. Accordingly, the subset of $\mathbf{Z} \times \mathbf{Z}$ for which f contains nonzero values is finite. The set of nonzero real values is denoted by $\Re - \{0\}$, and the subset of lattice points in $\mathbf{Z} \times \mathbf{Z}$ for which f takes on these values is given by the inverse function $f^{-1}(\Re - \{0\})$.

We define the *co-zero* set of f, $\text{coz}(f)$, to be $f^{-1}(\Re - \{0\})$. Thus, $\text{coz}(f)$ is the subset of $\mathbf{Z} \times \mathbf{Z}$ for which f does not have value zero. Any subset T of $\mathbf{Z} \times \mathbf{Z}$ for which $\text{coz}(f) \subset T$ is said to be a support region of f. Thus, f is zero outside of a support region consistent with the definition of this concept. Additionally, f is of finite support means that $\text{card}(T) < \infty$ for some support region T of f. As previously mentioned, an arbitrary support region of f in $\Re^{Z \times Z}$ can be denoted by $\text{supp}(f)$. Moreover, it follows that $\text{coz}(f)$ is the support region of minimal size.

The product of two functions f and g in $\Re^{Z \times Z}$ has a co-zero set equal to the intersection of the corresponding co-zero sets of the individual arguments, that is $\text{coz}(f \cdot g) = \text{coz}(f) \cap \text{coz}(g)$.

Another important feature of the co-zero set is seen when functions in $\Re^{Z \times Z}$ are to be added, only points in the union of their co-zero sets need to be considered. Function values outside these sets are zero. The following example illustrates the use of co-zero sets when multiplying and adding two functions in $\Re^{Z \times Z}$.

Example 3.1 *Let*

$$f = \begin{pmatrix} 3 & 2 \\ 0 & 1 \\ 2 & 0 \end{pmatrix}_{0,0}^{0} \quad \textit{and} \quad g = \begin{pmatrix} 3 & -1 & 0 \\ 0 & 1 & 0 \end{pmatrix}_{0,0}^{0}$$

It follows that

$$\text{coz}(f) = \{(0,0),(1,0),(1,-1),(0,-2)\}$$

$$\text{coz}(g) = \{(0,0),(1,0),(1,-1)\}$$

Therefore

$$\text{coz}(f) \cap \text{coz}(g) = \{(0,0),(1,0),(1,-1)\}$$

Since $f \cdot g = 0$ *outside* $\text{coz}(f) \cap \text{coz}(g)$, *to find* $f \cdot g$ *we only need to multiply the functions at the points* $(0,0),(1,0)$ *and* $(1,-1)$.

$$(f \cdot g)(0,0) = 3 \cdot 3 = 9$$

$$(f \cdot g)(1,0) = 2(-1) = -2$$

$$(f \cdot g)(1,-1) = 1 \cdot 1 = 1$$

So

$$f \cdot g = \begin{pmatrix} 9 & -2 \\ 0 & 1 \end{pmatrix}_{0,0}^{0}$$

$f + g$ *can be found by adding values* $f(i,j)$ *and* $g(i,j)$ *for* $(i,j) \in \text{coz}(f) \bigcup \text{coz}(g)$. *Doing this we obtain*

$$f + g = \begin{pmatrix} 6 & 1 \\ 0 & 2 \\ 2 & 0 \end{pmatrix}_{0,0}^{0}$$

The convolution of two digital signals f and g is formally given at each lattice point (n, m) by

$$(f * g)(n,m) = \sum_{k,j=-\infty}^{\infty} f(k,j) g(n-k, m-j)$$

If f and g are both of finite support, $f * g$ will also be of finite support. That is, the result will be zero except on some finite set, a support region of $f * g$.

Co-zero sets can be employed for the purpose of finding the support region for the convolution of the two signals. Suppose that f and g are of finite support, then a support region for $f * g$ is found by forming the dilation of coz(f) and coz(g). The dilation operation will be discussed in more detail in the Appendix.

The dilation of the co-zero set of f and g is found by forming the union of all the sets obtained from adding all points, a single point at a time, from coz(g) to every point in coz(f). That is,

$$D(\text{coz}(f), \text{coz}(g)) = \bigcup_{\substack{(k,j)\in\text{coz}(g)\\(n,m)\in\text{coz}(f)}} \{(n,m) + (k,j)\}$$

Clearly, $\text{coz}(f * g) \subset D(\text{coz}(f), \text{coz}(g))$.

Example 3.2 *Suppose*

$$f = \begin{pmatrix} 3 & 2 \\ 0 & 1 \\ 2 & 0 \end{pmatrix}_{0,0}^{0} \quad \textit{and} \quad g = \begin{pmatrix} 3 & -1 & 0 \\ 0 & 1 & 0 \end{pmatrix}_{0,0}^{0}$$

as in Example 3.1. *Then* $\text{coz}(f) = \{(0,0),(1,0),(1,-1),(0,-2)\}$ *and* $\text{coz}(g) = \{(0,0),(1,0),(1,-1)\}$. *We first add to every point in* coz(f) *the point* $(0,0)$, *then* $(1,0)$ *and finally* $(1,-1)$. *Then union them together to obtain*

$$D(\text{coz}(f), \text{coz}(g)) =$$

$$\{(0,0),(1,0),(1,-1),(0,-2),(2,0),(2,-1),(1,-2),(2,-2),(1,-3)\}$$

So the convolution, which will be found in the next section, will surely be zero outside $D(\text{coz}(f), \text{coz}(g))$. It may also be zero for some points in this set. Thus, the computation just given provides a support region for the convolution.

3.2 Convolution for Signals of Finite Support

As mentioned in the previous section, the convolution of two signals of finite support will also be of finite support. Thus $f * g$ will be zero outside a finite support region of $f * g$.

Consider the subset F of $\Re^{Z\times Z}$, being the set of all functions of finite support. The *convolution* $f * g$ of two functions f and g in this space is defined using the pointwise definition:

$$(f * g)(n,m) = \sum_{\substack{(k,j)\in\mathrm{coz}(f)\\ (n-k,m-j)\in\mathrm{coz}(g)}} f(k,j)g(n-k,m-j)$$

This formula gives the same answer as the convolution formula given in the last section. However, the above formula involves a minimal number of computations. Since the dilation of $\mathrm{coz}(f)$ with $\mathrm{coz}(g)$ is a support region for $f * g$, the convolution should be found by first determining $D(\mathrm{coz}(f), \mathrm{coz}(g))$ and then using

$$(f * g)(n,m) =$$

$$\begin{cases} \displaystyle\sum_{\substack{(k,j)\in\mathrm{coz}(f)\\ (n-k,m-j)\in\mathrm{coz}(g)}} f(k,j)g(n-k,m-j) & (n,m) \in D(\mathrm{coz}(f), \mathrm{coz}(g)) \\ 0 & \mathit{otherwise} \end{cases}$$

Example 3.3 *Let f and g be given using the bound matrices*

$$f = \begin{pmatrix} 3 & 0 \\ 2 & -1 \end{pmatrix}_{1,2}^{0} \quad \mathit{and} \quad g = \begin{pmatrix} -2 & 1 \\ 0 & 3 \end{pmatrix}_{3,4}^{0}$$

These functions are illustrated in Figures 3.1(a) *and* (b), *respectively.*

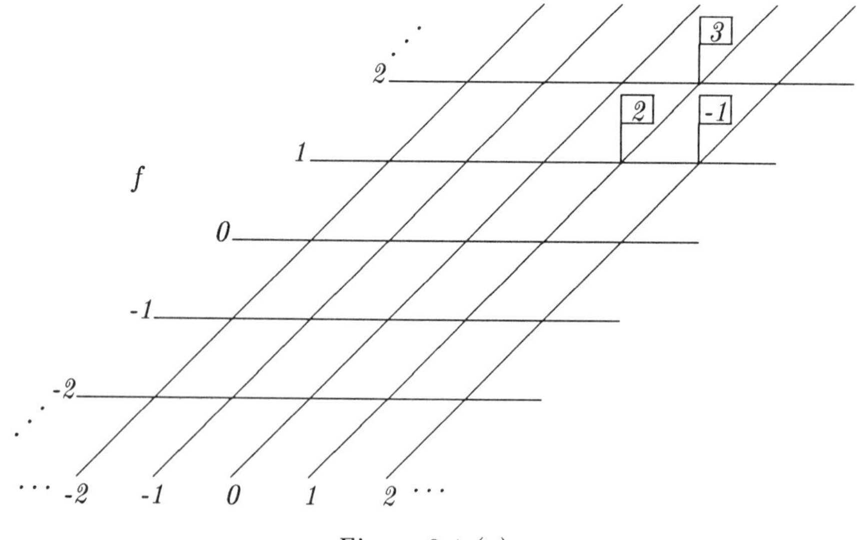

Figure 3.1.(a)

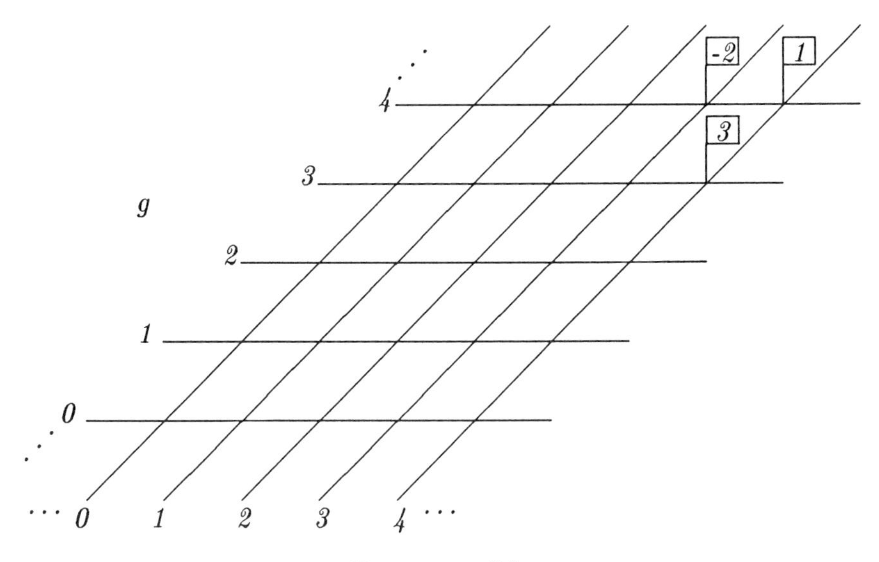

Figure 3.1.(b)

In order to find $f * g$, *we first find the co-zero sets of* f *and* g

$$\operatorname{coz}(f) = \{(1,2),(1,1),(2,1)\}$$

$$\operatorname{coz}(g) = \{(3,4),(4,4),(4,3)\}$$

Next, we find the dilation of $\operatorname{coz}(f)$ *and* $\operatorname{coz}(g)$,

$$D(\operatorname{coz}(f),\operatorname{coz}(g)) = \{(4,6),(4,5),(5,5),(5,6),(6,5)(5,4),(6,4)\}$$

The convolution $f * g$ *is zero outside of this set. We will now determine* $f * g$ *for points in this set. To do this we choose points* (n,m) *in* $D(\operatorname{coz}(f),\operatorname{coz}(g))$ *one at a time. For each point, so chosen, we then choose points* (k,j) *in* $\operatorname{coz}(f)$. *If the point* $(n-k,m-j)$ *is also in* $\operatorname{coz}(g)$ *we form the product* $f(k,j)g(n-k,m-j)$. *For instance, choose* $(n,m) = (4,6)$, *and take*

$$(k,j) = (1,2) \in \operatorname{coz}(f)$$

then we see if

$$(n,m) - (k,j) = (3,4)$$

is in $\operatorname{coz}(g)$. *Sure it is, then we form*

$$f(1,2) \cdot g(3,4) = 3(-2) = -6$$

Next, we take $(1,1) \in \mathrm{coz}(f)$. *This time*

$$(n,m) - (k,j) = (3,5)$$

is not in $\mathrm{coz}(g)$ *and the product is not formed. Also, for* $(2,1) \in \mathrm{coz}(f)$, $(n,m) - (2,1) = (2,5) \notin \mathrm{coz}(g)$. *Hence*

$$(f * g)(4,6) = f(1,2)g(3,4) = -6.$$

Similarly, we have

$$(f * g)(4,5) = f(1,1)g(3,4) = -4$$
$$(f * g)(5,5) = f(1,2)g(3,4) + f(1,1)g(4,4) + f(2,1)g(3,4) = 13$$
$$(f * g)(5,6) = f(1,2)g(4,4) = 3$$
$$(f * g)(6,5) = f(2,1)g(4,4) = -1$$
$$(f * g)(5,4) = f(1,1)g(4,3) = 6$$
$$(f * g)(6,4) = f(2,1)g(4,3) = -3$$

Consequently,

$$f * g = \begin{pmatrix} -6 & 3 & 0 \\ -4 & 13 & -1 \\ 0 & 6 & -3 \end{pmatrix}_{4,6}^{0}$$

This function is illustrated in Figure 3.2.

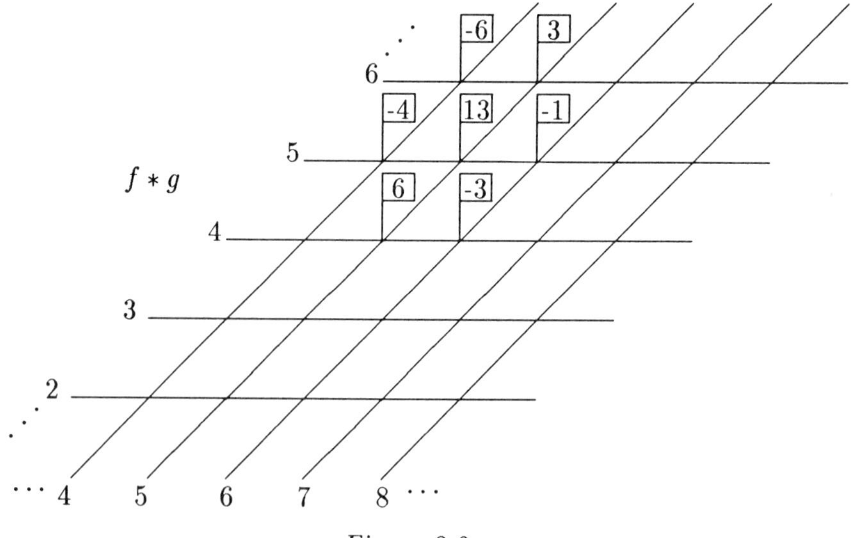

Figure 3.2

An equivalent way of writing the convolution is

$$(f * g)(n,m) = \sum_{(k,j)\in \operatorname{coz}(f)\cap(-\operatorname{coz}(g)+(n,m))} f(k,j)g(n-k,m-j)$$

For instance, in Example 3.3 observe that

$$(f * g)(6,4) = \sum_{(k,j)\in \operatorname{coz}(f)\cap(-\operatorname{coz}(g)+(6,4))} f(k,j)g(6-k,4-j)$$

This expression holds true by referring to the Appendix, where it is seen that

$$-\operatorname{coz}(g) = \{(-3,-4),(-4,-4),(-4,-3)\}$$

Hence

$$-\operatorname{coz}(g) + (6,4) = \{(3,0),(2,0),(2,1)\}$$

and

$$(k,j) \in \operatorname{coz}(f)\bigcap(-\operatorname{coz}(g)+(6,4)) = \{(2,1)\}$$

then it follows that

$$(f * g)(6,4) = f(2,1)g(4,3) = -3$$

Similarly, we can continue this process

$$(f * g)(6,5) = \sum_{(k,j)\in \operatorname{coz}(f)\bigcap(-\operatorname{coz}(g)+(6,5))} f(k,j)g(6-k,5-j)$$

3.3 Bound Matrices for Convolution

The convolution of two functions f and g, of finite support in $\Re^{Z\times Z}$, can be found more simply by making greater use of operations involving bound matrices. A nonzero function in the aforementioned class is represented uniquely by its minimal bound matrix. This gives a compact notation as well as geometric insight. Suppose that the signal f is given using its minimal $m\times n$ bound matrix representation

$$f = \begin{pmatrix} a_{11} & a_{12} & \cdots & a_{1n} \\ a_{21} & a_{22} & \cdots & a_{2n} \\ \cdot & \cdot & & \\ \cdot & \cdot & & \\ \cdot & \cdot & & \\ a_{m1} & a_{m2} & \cdots & a_{mn} \end{pmatrix}_{p,q}^{0}$$

That is, at least one a_{ij} is nonzero for $i = 1$, $i = m$, $j = 1$ and $j = n$.

The objective is to form the convolution of f and g, where g is also given as a minimal bound matrix. Thus,

$$g = \begin{pmatrix} b_{11} & b_{12} & \cdots & b_{1s} \\ b_{21} & b_{22} & \cdots & b_{2s} \\ \cdot & \cdot & & \\ \cdot & \cdot & & \\ \cdot & \cdot & & \\ b_{r1} & b_{r2} & \cdots & b_{rs} \end{pmatrix}_{u,v}^{0}$$

where, as in the bound matrix for f, at least one b_{ij} is nonzero on "the boundary".

The convolution $f * g$ will be represented by a bound matrix with $(r + m - 1)(s + n - 1)$ entries, the support region of $f * g$ will be rectangular with upper left corner $(p + u, q + v)$. Several steps are needed in obtaining $f * g$. First, we form the 180° rotation of g, thereby obtaining g', such that $g'(k, j) = g(-k, -j)$. Using minimal bound matrix representation, notice that

$$g' = \begin{pmatrix} b_{rs} & b_{r,s-1} & \cdots & b_{r1} \\ b_{r-1,s} & b_{r-1,s-1} & \cdots & b_{r-1,1} \\ \cdot & \cdot & & \\ \cdot & \cdot & & \\ \cdot & \cdot & & \\ b_{1s} & b_{1,s-1} & \cdots & b_{11} \end{pmatrix}_{-u-s+1,-v+r-1}^{0}$$

The second step is to pad the minimal representation of f with $s - 1$ columns of zeros to the left and to the right, and $r - 1$ rows of zeros above and below.

The next several steps are used in obtaining a total number of $(n + s - 1)(m + r - 1)$ translates of g'. The minimal bound matrix

g' should be translated such that it is wholly contained inside the padded bound matrix of f. For each translation index such that this is true, a dot or inner product of the a_{ij}'s and the b_{kl}'s located at he same point in the integral lattice are found, and the sum of all such products taken. The resulting value defines the convolution $f * g$ at the index of translation. An example should make this clearer.

Example 3.4 *Using*

$$f = \begin{pmatrix} 3 & 0 \\ 2 & -1 \end{pmatrix}^{0}_{1,2} \quad \text{and } g = \begin{pmatrix} -2 & 1 \\ 0 & 3 \end{pmatrix}^{0}_{3,4}$$

as given in Example 3.3, *we will use the method stated above to find* $f * g$. *Since* g *is a two by two bound matrix, we pad the minimal bound matrix representation for* f *with one row above, below, left and right. That is, we write* f *as*

$$f = \begin{pmatrix} 0 & 0 & 0 & 0 \\ 0 & 3 & 0 & 0 \\ 0 & 2 & -1 & 0 \\ 0 & 0 & 0 & 0 \end{pmatrix}^{0}_{0,3}$$

Next we form the 180° *rotation of* g, *i.e., we input* g *into the block diagram*

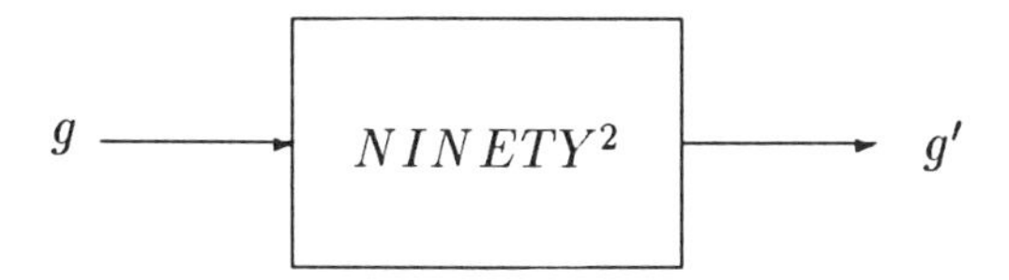

and get

$$g' = \begin{pmatrix} 3 & 0 \\ 1 & -2 \end{pmatrix}^{0}_{-4,-3}$$

Finally, we translate g' *such that it is located at any possible place within the "padded" bound matrix representation* f. *To begin the translation procedure we form*

$$g'_{4,6} = \begin{pmatrix} 3 & 0 \\ 1 & -2 \end{pmatrix}^{0}_{0,3}$$

and $(f * g)(4,6)$ *is found by forming the dot product of*

$$\begin{pmatrix} 3 & 0 \\ 1 & -2 \end{pmatrix}^{0}_{0,3} \quad \textit{with} \quad \begin{pmatrix} 0 & 0 \\ 0 & 3 \end{pmatrix}^{0}_{0,3}$$

This yields a value of -6. *Next, form the eight additional translates of* g

$$g'_{4,5} = \begin{pmatrix} 3 & 0 \\ 1 & -2 \end{pmatrix}^{0}_{0,2}$$

an $(f * g)(4,5)$ *is found by taking the dot product of*

$$\begin{pmatrix} 3 & 0 \\ 1 & -2 \end{pmatrix}^{0}_{0,2} \quad \textit{with} \quad \begin{pmatrix} 0 & 3 \\ 0 & 2 \end{pmatrix}^{0}_{0,2}$$

thereby yielding -4. *Continue to use the same method, finally we have*

$$f * g = \begin{pmatrix} -6 & 3 & 0 \\ -4 & 13 & -1 \\ 0 & 6 & -3 \end{pmatrix}^{0}_{4,6}$$

3.4 Parallel Convolution Algorithms

The convolution of two dimensional signals f and g can also be found using a convolution machine. Unlike the methods used in sections 3.2 and 3.3, this algorithm does not utilize pointwise operations.

The parallel convolution algorithm presented herein is specified in terms of several range and domain induced operations previously defined. Notice that

$$f * g = \sum_{(n,m) \in \mathrm{coz}(g)} [f_{n,m} \cdot g(n,m)]$$

and the formula can be equivalently written as

$$f * g = ADD_{(n,m)\in \mathrm{coz}(g)}[SCALAR(TRAN(f; n, m), g(n,m))]$$

The block diagram illustrating this implementation of convolution is given in Figure 3.3.

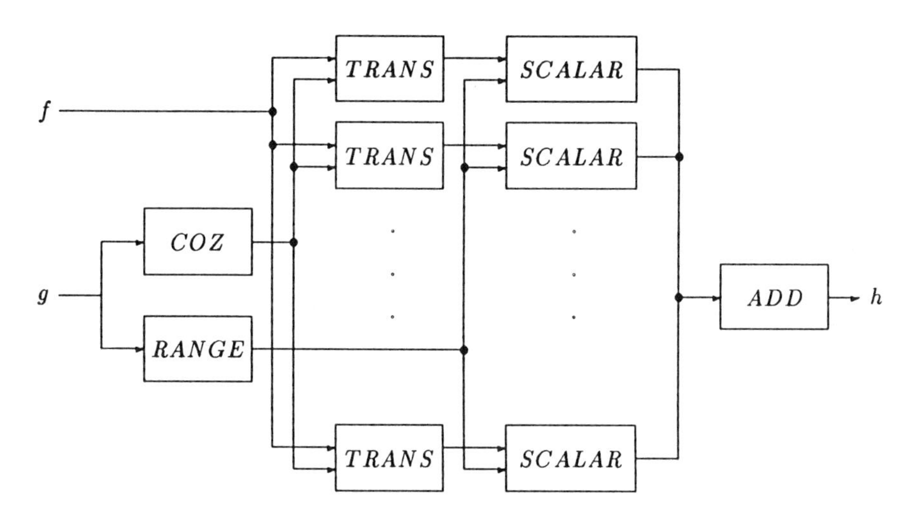

Figure 3.3
Parallel Convolution Machine

In this diagram we see the $TRAN$, $SCALAR$ and ADD blocks as described in chapter 2. Moreover, two new blocks appear. They are the co-zero set block, denoted by

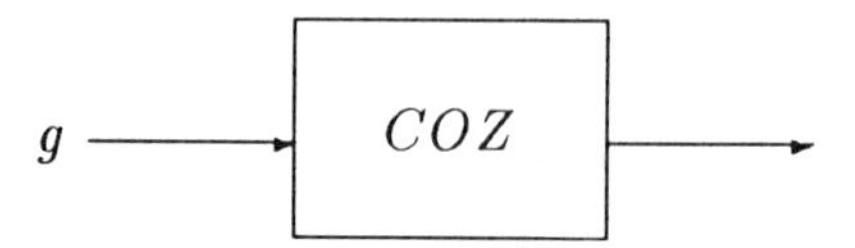

and the range of the co-zero set block, denoted by

The COZ block determines the set $\text{coz}(g)$, and sends each point in this set to a distinct $TRAN$ block. Thus, each $TRAN$ block is assigned a unique point from $\text{coz}(g)$. This two dimensional point is used in determining how much f should be translated. A constructive procedure for implementing this is to place the points in $\text{coz}(g)$ into a stack and pop the stack one at a time into each $TRAN$ block. Concomitantly, another stack implementing the $RANGE$ block extracts the corresponding value of g for each point in $\text{coz}(g)$, the $g(n, m)$ in the stack is popped into the $SCALAR$ block. For each point (n, m) in $\text{coz}(g)$, the value $g(n, m)$ is put into the same $SCALAR$ block for which $TRAN(f; n, m)$ is the other input.

The output of all the $SCALAR$ blocks are then added together to give the result $f * g$. The order in which the points and the corresponding values of g are input into the stack does not matter, what is of consequence is that when both stacks are popped simultaneously the function g arises. Throughout the text, the points and values are input using column domination.

Example 3.5 *As in Example 3.4, let*

$$f = \begin{pmatrix} 3 & 0 \\ 2 & -1 \end{pmatrix}^{0}_{1,2} \quad \textit{and} \quad g = \begin{pmatrix} -2 & 1 \\ 0 & 3 \end{pmatrix}^{0}_{3,4}$$

we will find the convolution of f and g using the parallel algorithm. Accordingly, let the output of the COZ block be the elements in $\mathrm{coz}(g)$ *arranged in a stack. Also, let the output of the $RANGE$ block be the values of g arranged in a stack in the same order as the corresponding elements of* $\mathrm{coz}(g)$.

$$f * g = \begin{pmatrix} -6 & 0 & 0 \\ -4 & 2 & 0 \\ 0 & 0 & 0 \end{pmatrix}^{0}_{4,6} + \begin{pmatrix} 0 & 3 & 0 \\ 0 & 2 & -1 \\ 0 & 0 & 0 \end{pmatrix}^{0}_{4,6} + \begin{pmatrix} 0 & 0 & 0 \\ 0 & 9 & 0 \\ 0 & 6 & -3 \end{pmatrix}^{0}_{4,6}$$

$$= \begin{pmatrix} -6 & 3 & 0 \\ -4 & 13 & -1 \\ 0 & 6 & -3 \end{pmatrix}^{0}_{4,6}$$

Consult Figure 3.4 *for an execution trace in the solution.*

In the Figure 3.4, notice that we write

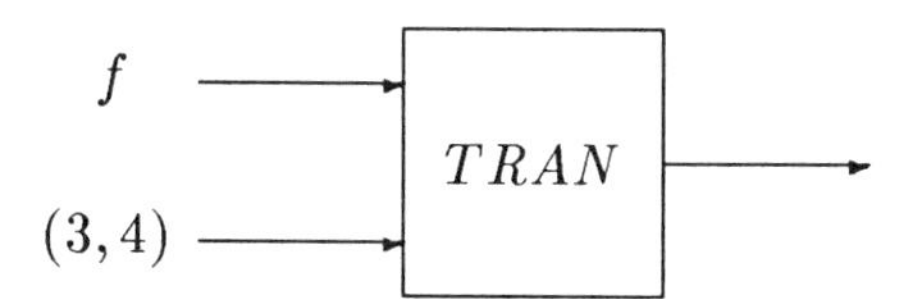

instead of the previously defined notation given for $TRAN$. Both notations result in the same output, $f_{3,4}$. Additionally, anticipating the final addition, we should always write all the outputs of the $TRAN - SCALAR$ arrangements using a common size, common location bound matrix. For instance, we write

$$\begin{pmatrix} -6 & 0 & 0 \\ -4 & 2 & 0 \\ 0 & 0 & 0 \end{pmatrix}^{0}_{4,6}$$

instead of

$$\begin{pmatrix} -6 & 0 \\ -4 & 2 \end{pmatrix}^{0}_{4,6}$$

The same representation should be used for the output of all other $TRAN - SCALAR$ arrangements.

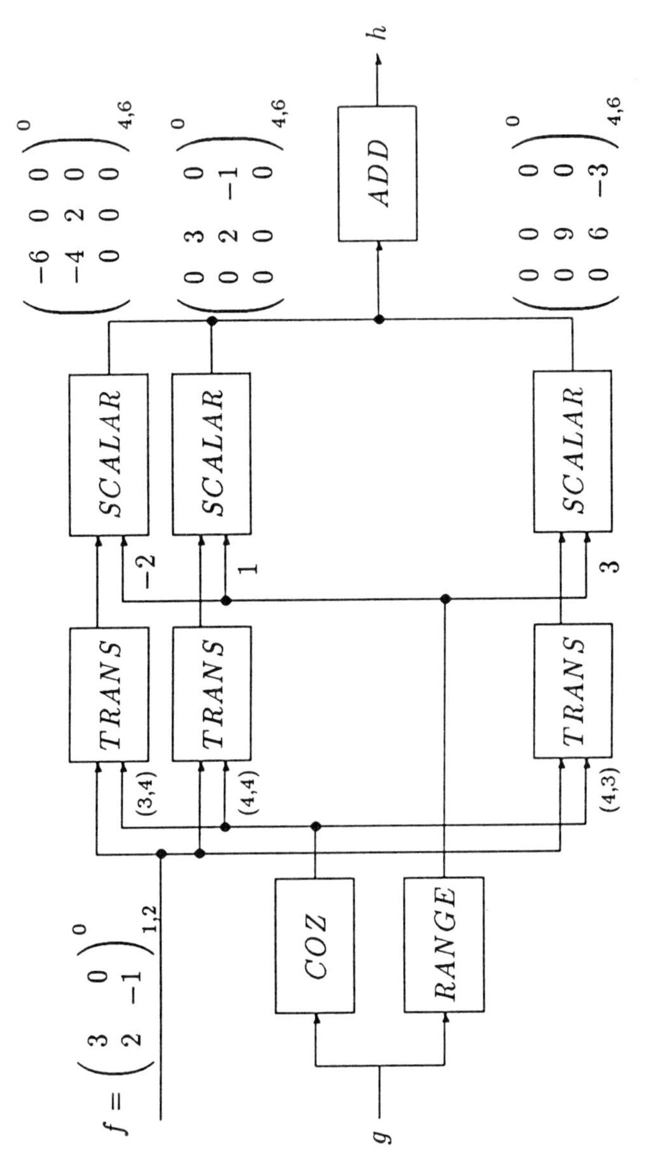

$f = \begin{pmatrix} 3 & 0 \\ 2 & -1 \end{pmatrix}^{0}_{1,2}$
g
COZ
RANGE
TRANS
TRANS
TRANS
(3,4)
(4,4)
(4,3)
SCALAR
SCALAR
SCALAR
−2
1
3
ADD
h
$\begin{pmatrix} -6 & 0 & 0 \\ -4 & 2 & 0 \\ 0 & 0 & 0 \end{pmatrix}^{0}_{4,6}$
$\begin{pmatrix} 0 & 3 & 0 \\ 0 & 2 & -1 \\ 0 & 0 & 0 \end{pmatrix}^{0}_{4,6}$
$\begin{pmatrix} 0 & 0 & 0 \\ 0 & 9 & 0 \\ 0 & 6 & -3 \end{pmatrix}^{0}_{4,6}$

Figure 3.4

3.5 Convolution of Signals of Non-Finite Support

Consider the convolution of the digital signals, f and g which are not necessarily of finite support. Since these signals might have infinite extents we must worry whether or not the convolution exists. That is, it is not certain whether or not the series

$$\sum_{k,j=-\infty}^{\infty} f(k,j)g(n-k,m-j)$$

converges.

The convolution always exist if one of f or g is of finite support. For instance, f is an arbitrarily two dimensional signal and g is of finite support. In this case, $\text{card}(\text{coz}(g)) < \infty$ and for any point (n, m), the equation

$$(f * g)(n,m) = \sum_{k,j=-\infty}^{\infty} f(k,j)g(n-k,m-j)$$

is always a finite sum involving no more than $\text{card}(\text{coz}(g))$ additions. Accordingly, each of the methods given in sections 3.2, 3.3 and 3.4 for determining $f * g$ can be slightly modified for this situation. As a matter of fact, the parallel method is most easily describable. It involves precisely the same figure, namely Figure 3.3. In this case, f is given as an infinite bound matrix and all subsequent calculations are fully applicable. For instance, all the operations, $TRAN$, $SCALAR$ and ADD, when used a finite number of times can involve arbitrary signals in $\Re^{Z \times Z}$. Thus, Figure 3.3 is applicable as given. We will provide an illustration of the use of this algorithm, for f and g, when f is not of finite support and $\text{card}(\text{coz}(g)) < \infty$.

Example 3.6 *Suppose that f is the first quadrant step function u,*

$$f = \begin{pmatrix} \cdot & \cdot & \\ \cdot & \cdot & \\ \cdot & \cdot & \\ 1 & 1 & \cdots \\ \boxed{1} & 1 & \cdots \end{pmatrix}^{0}$$

and

$$g = \begin{pmatrix} 2 & 0 \\ \boxed{-1} & 3 \end{pmatrix}^0$$

Inputting g through the COZ block, we obtain

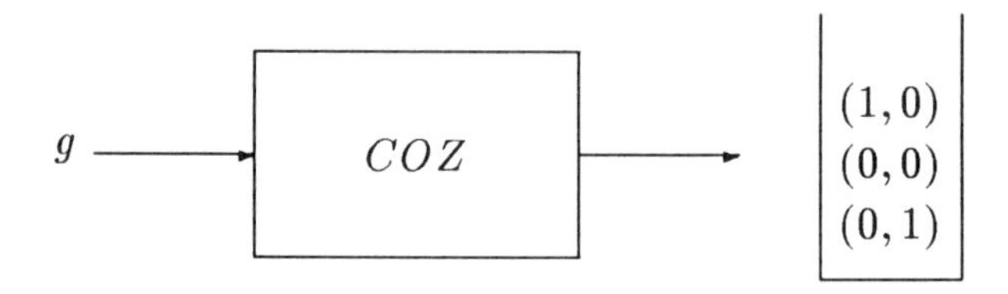

and through the RANGE block

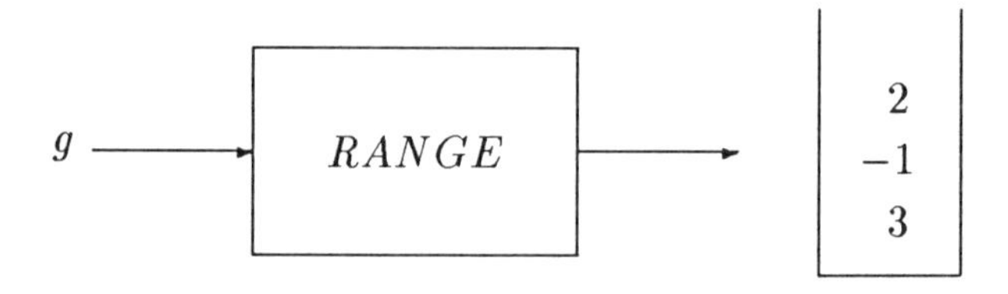

Referring to Figure 3.5, we have

$$SCALAR(TRAN(f;0,1);2) = 2u_{0,1}$$

$$= \begin{pmatrix} \cdot & \cdot & \cdot & \\ \cdot & \cdot & \cdot & \\ \cdot & \cdot & \cdot & \\ 2 & 2 & 2 & \cdots \\ 2 & 2 & 2 & \cdots \\ \boxed{0} & 0 & 0 & \cdots \end{pmatrix}^0$$

$$SCALAR(TRAN(f;0,0);-1) = -u$$

$$= \begin{pmatrix} \cdot & \cdot & \cdot & \\ \cdot & \cdot & \cdot & \\ \cdot & \cdot & \cdot & \\ -1 & -1 & -1 & \cdots \\ -1 & -1 & -1 & \cdots \\ \boxed{-1} & -1 & -1 & \cdots \end{pmatrix}^0$$

and

$$SCALAR(TRAN(f;1,0);3) = 3u_{1,0}$$

$$= \begin{pmatrix} \cdot & \cdot & \cdot & \\ \cdot & \cdot & \cdot & \\ \cdot & \cdot & \cdot & \\ 0 & 3 & 3 & \cdots \\ 0 & 3 & 3 & \cdots \\ \boxed{0} & 3 & 3 & \cdots \end{pmatrix}^{0}$$

Therefore

$$f * g = 2u_{0,1} - u + 3u_{1,0}$$

$$= \begin{pmatrix} \cdot & \cdot & \cdot & \\ \cdot & \cdot & \cdot & \\ \cdot & \cdot & \cdot & \\ 1 & 4 & 4 & \cdots \\ 1 & 4 & 4 & \cdots \\ \boxed{-1} & 2 & 2 & \cdots \end{pmatrix}^{0}$$

There are numerus sufficient conditions for the existence of $f * g$ when both f and g are not of finite support. Among the simplest, and yet one of the most important convergence criteria occurs when f and g both belong to the space l_1. That is, the sum of the absolute value of $f(n,m)$ and $g(n,m)$ are convergent. By definition, f and g belong to the space l_1 if:

$$\sum_{m,n=-\infty}^{\infty} | f(n,m) | < \infty$$

and

$$\sum_{m,n=-\infty}^{\infty} | g(n,m) | < \infty$$

The order of summation is irrelevant for functions in l_1. Example 3.7 will illustrate this fact.

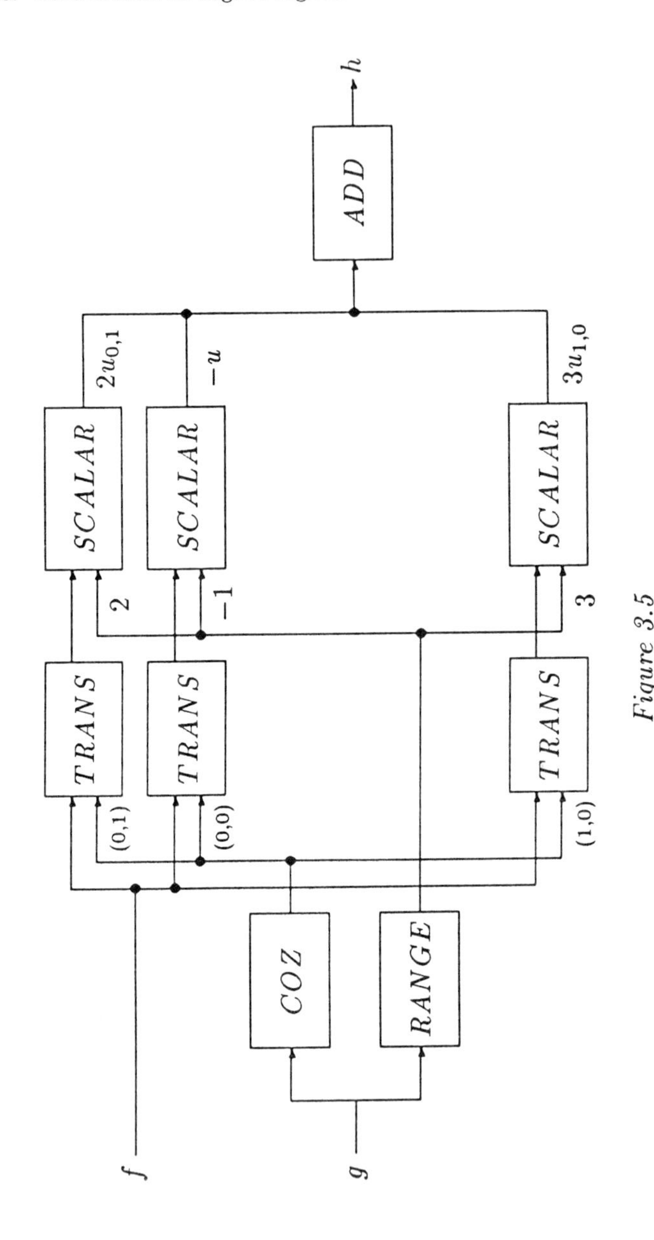

f
g
COZ
RANGE
TRANS
TRANS
TRANS
(0,1)
(0,0)
(1,0)
SCALAR
SCALAR
SCALAR
2
−1
3
$2u_{0,1}$
$-u$
$3u_{1,0}$
ADD
h

Figure 3.5

Example 3.7 *The function*

$$f = \begin{pmatrix} \cdot & & & & \\ \cdot & & & & \\ \cdot & & & & \\ 1/2^6 & \cdots & & & \\ 1/2^5 & 1/2^7 & \cdots & & \\ 1/2 & 1/2^4 & 1/2^8 & \cdots & \\ \boxed{1} & 1/2^2 & 1/2^3 & 1/2^9 & \cdots \end{pmatrix}$$

is in l_1. *Indeed, we have the sum*

$$\sum_{m,n=-\infty}^{\infty} f(n,m) = 2$$

If f and g are in l_1 then so are $f+g$ and af. Hence, the space l_1 is itself a vector space under addition and scalar multiplication. Accordingly, l_1 is a subspace of $\Re^{Z \times Z}$. Moreover, every digital signal of finite support is in l_1, and so the vector space of digital signals of finite support is a subspace of l_1.

When f and g are in l_1 then the convolution exists, it is given for each lattice point (n,m) by

$$f * g = \sum_{k,j=-\infty}^{\infty} f(k,j)g(n-k, m-j)$$

Example 3.8 *Suppose* f *and* g *are given as follows:*

$$f = \begin{pmatrix} 1 & 1/2 & 1/2^2 & 1/2^3 & \cdots \end{pmatrix}_{0,0}^{0}$$

$$g = \begin{pmatrix} \cdot \\ \cdot \\ \cdot \\ 1/2^2 \\ 1/2 \\ \boxed{1} \end{pmatrix}^{0}$$

then

$$f * g = \begin{pmatrix} \cdot & \cdot & \cdot & \cdot & \\ \cdot & \cdot & \cdot & \cdot & \\ \cdot & \cdot & \cdot & \cdot & \\ 1/2^2 & 1/2^3 & 1/2^4 & 1/2^5 & \cdots \\ 1/2 & 1/2^2 & 1/2^3 & 1/2^4 & \cdots \\ \boxed{1} & 1/2 & 1/2^2 & 1/2^3 & \cdots \end{pmatrix}^0$$

This follows since

$$f(k,j) = 0 \text{ for all } j \neq 0 \text{ or } k < 0$$

and

$$g(n-k, m-j) = 0 \text{ for all } n-k \neq 0 \text{ or } m-j < 0.$$

For $j = 0$ *and* $k \geq 0$ *we have*

$$f(k,j) = f(k,0) = 1/2^k$$

For $n - k = 0$ *and* $m - j \geq 0$,

$$g(n-k, m-j) = g(0, m-j) = 1/2^{m-j}.$$

So

$$(f*g)(n,m) = \sum_{k,j=-\infty}^{\infty} f(k,j)g(n-k, m-j) = f(n,0)g(0,m) = 1/2^{n+m}$$

for $n, m \geq 0$.

A very important property of the convolution is that the function $f * g$ is also in l_1. That is,

$$\sum_{n,m=-\infty}^{\infty} |(f*g)(n,m)| < \infty$$

which implies

$$* : l_1 \times l_1 \to l_1$$

Example 3.9 *Referring to Example 3.8, we have*

$$(f * g)(n, m) = 1/2^{n+m}$$

for $n, m \geq 0$. *Therefore*

$$\sum_{n,m=-\infty}^{\infty} |(f * g)(n, m)| = 4$$

The convolution operation for signals in l_1 (which includes signals of finite support) will be denoted by the following block diagram

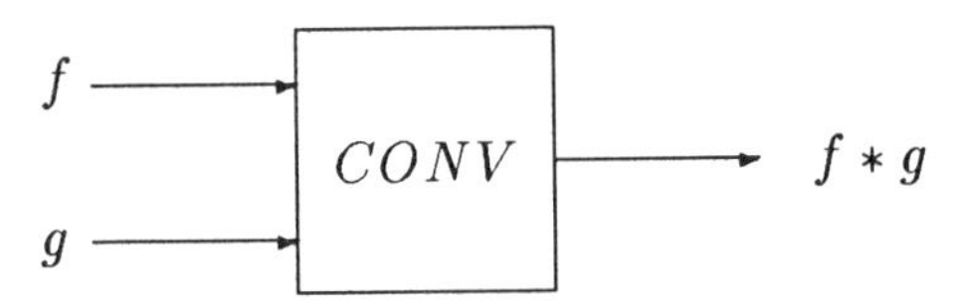

3.6 Banach Algebra Properties of Convolution in l_1

Every digital signal of finite support is in l_1. Consequently, we will describe the properties of convolution in l_1, these properties necessarily hold for signals of finite support.

If f, g and h all are in l_1 and a is any real number, then the following hold:

B1) Associative Law: $f * (g * h) = (f * g) * h$

B2) Distributive Laws:

a) $f * (g + h) = f * g + f * h$

b) $(f + g) * h = f * h + g * h$

B3) Commutative Law: $f * g = g * f$

B4) Identity (Unity) Law: $f * \delta = \delta * f = f$

B5) Scalar Commutativity: $a(f * g) = (af) * g = f * (ag)$

The following block digrams demonstrate how the above laws can be applied. Because of the associative law, given by property **B1)**, we can use the block diagram

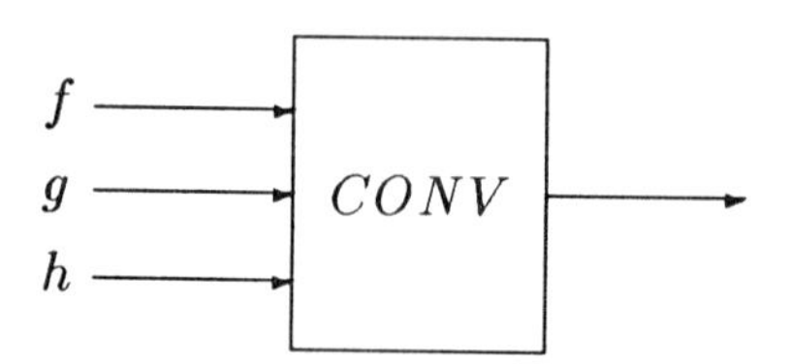

instead of

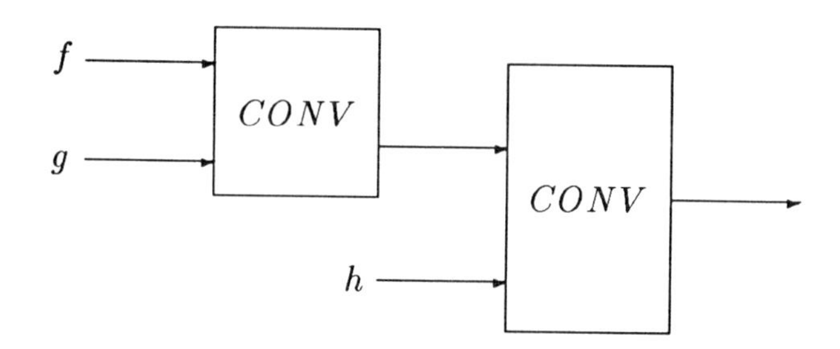

or

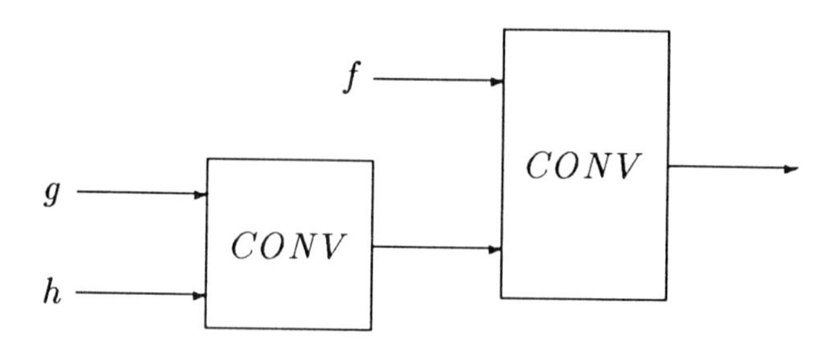

Block diagrams illustrating the first distributive law, given in property **B2)**, are provided below. The output of

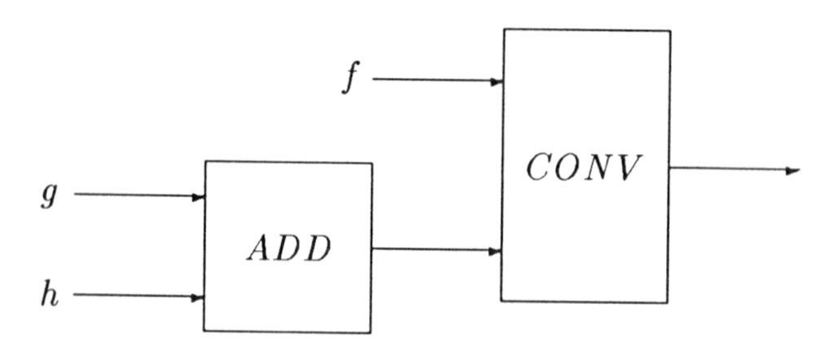

and

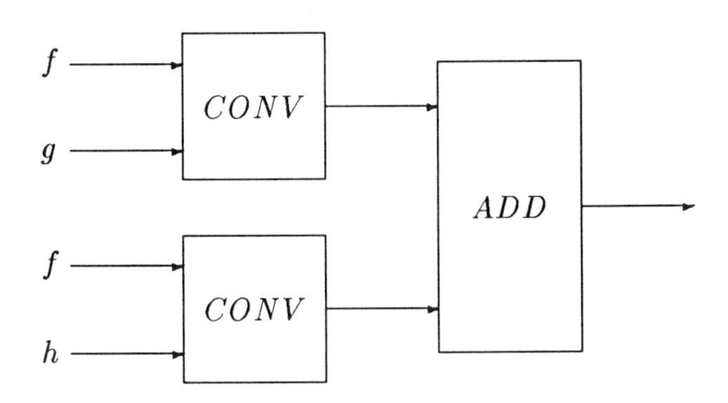

are equal.

Property **B5)** requires that the output of the following three block diagrams are all equal:

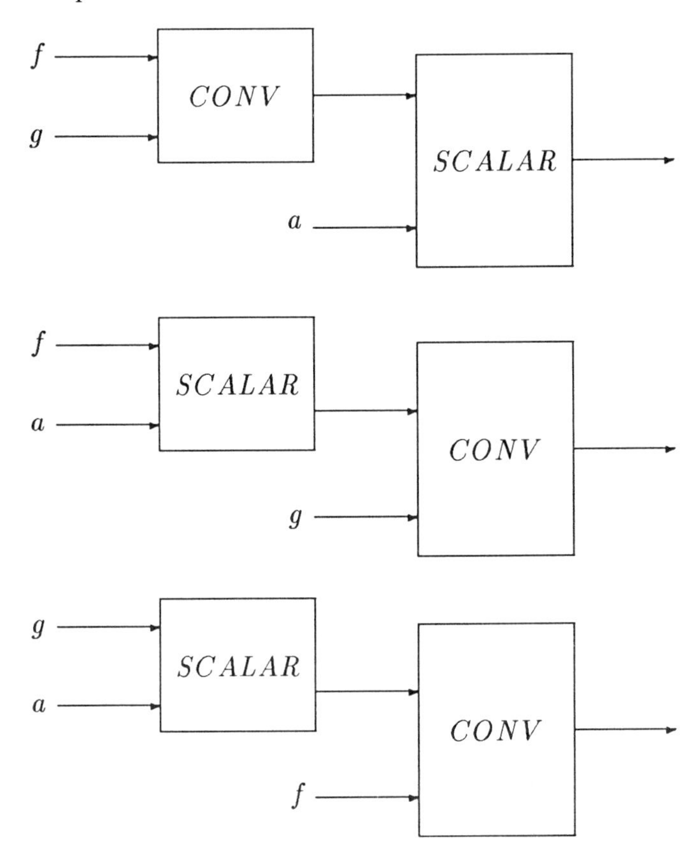

The strength of signals in l_1 is often measured using the one norm $\|\cdot\|_1$ where

$$\|\cdot\|_1 : l_1 \to R$$

and

$$\|f\|_1 = \sum_{n,m=-\infty}^{\infty} | f(n,m) |$$

The following properties are satisfied for f and g in l_1 and any real number a.

N1) Positive definite: $\|f\|_1 \geq 0$ and $\|f\|_1 = 0$ if $f = 0$
N2) Homogenous: $\|af\|_1 = |a| \, \|f\|_1$
N3) Triangle Inequality: $\|f + g\|_1 \leq \|f\|_1 + \|g\|_1$
N4) Convolutional Triangle Inequality: $\|f * g\|_1 \leq \|f\|_1 \cdot \|g\|_1$

Example 3.10 *Consider the digital signals f and g,*

$$f = \begin{pmatrix} 3 & 2 & 1 \end{pmatrix}_{0,0}^{0} \quad \text{and} \quad g = \begin{pmatrix} 1 & 0 \\ 2 & -1 \end{pmatrix}_{0,0}^{0}$$

$$\|f\|_1 = \left\| \begin{pmatrix} 3 & 2 & 1 \end{pmatrix}_{0,0}^{0} \right\|_1 = 6$$

and

$$\|g\|_1 = \left\| \begin{pmatrix} 1 & 0 \\ 2 & -1 \end{pmatrix}_{0,0}^{0} \right\|_1 = 4$$

Computing the convolution yields

$$f * g = \begin{pmatrix} 3 & 2 & 1 & 0 \\ 6 & 1 & 0 & -1 \end{pmatrix}_{0,0}^{0}$$

Notice that

$$\|f * g\|_1 = 14$$

and

$$\|f * g\|_1 \leq \|f\|_1 \cdot \|g\|_1$$

Properties **N1)**, **N2)** and **N3)** together with the vector space structure shows that l_1 is a normed vector space. If **N4)** along with **B1)** through **B5)** are employed, l_1 has all the algebraic structure of a commutative Banach algebra with identity.

3.7 Filtering by Convolution

Convolution is mainly performed for the purpose of shaping or filtering a given signal. Here, one of the signals, say f, is obtained empirically by collecting data or utilizing sensors. The signal f, consequently, is called an observed digital signal. The other signal, which we shall denote by h, is obtained in a heuristic fashion. It must be known, a priori, that the application of h to f results in a new digital signal with more desirable characteristics than f originally possessed. The latter signal h is called an impulse response or more suggestively, a structured signal.

One reason for performing the convolution of f with h might be for restoration purposes. Here, f is obtained through the use of sensors and h acts like an inverse model for the sensor in question. The convolution produces a function with less sensor distortion than f. Another reason for convolving f with h might be for signal enhancement. In this case, we purposely distort f by convolving it with h in order to simplify certain characteristics in f and to attenuate other characteristics. In all these cases, h is most often subjectively chosen.

Moving average filters provide a simple, yet useful way of performing certain types of signal restoration and enhancement. The output signal at a given point is equal to a finite weighted average of values of the input signal. By an average we mean the weights are nonnegative and sum to one. The values used in the averaging procedure, more often than not, are close in proximity to the point at which the calculation is to be performed. Thus, a moving average is often classified as a local or neighborhood procedure.

A neighborhood of lattice points is a set consisting of the point itself along with the eight vertically, horizontally, and diagonally adjacent neighbors. See Figure 3.6(a). The strong neighborhood of lattice points is the point itself along with just the two horizontal and two vertical adjacent neighbors. See Figure 3.6(b).

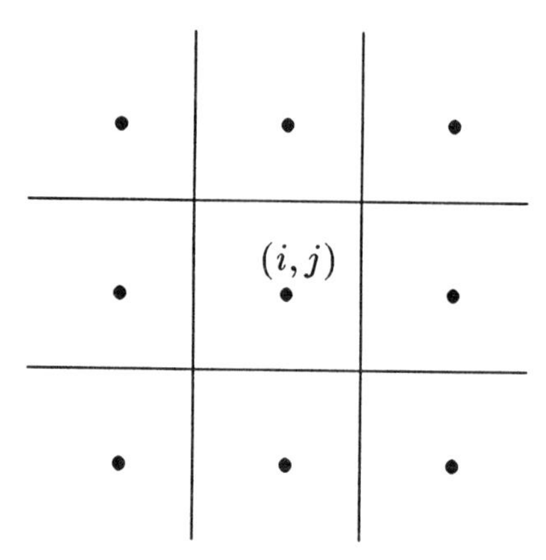

Figure 3.6(a)

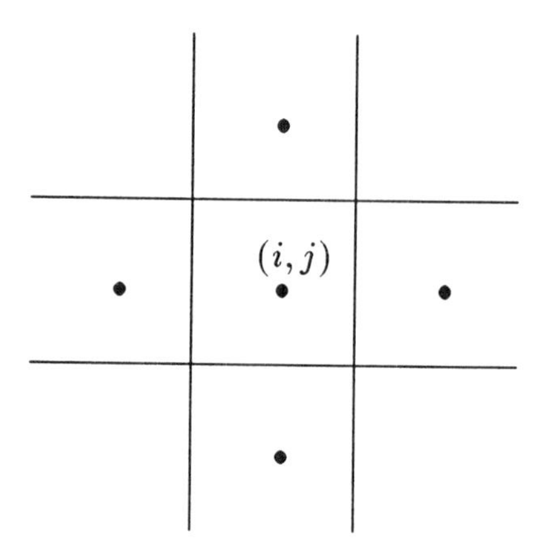

Figure 3.6(b)

A moving average filter yields a value for the output signal at a point (p, q). This value often involves values of the input signal at points in the neighborhood of (p, q). Such a filter also smooths out the values in a signal. It makes the output more uniform and thereby less varying. Accordingly, it is a type of low-pass filter. When applied to a constant signal the output will equal the input, and if the input is "almost" constant, the output will be even more constant.

Example 3.11 *Suppose that the following moving average filter is defined by the input/output relationship:*

$$g(n,m) = \frac{4f(n,m) + 3f(n-1,m) + 2f(n,m-1) + f(n-1,m-1)}{10}$$

In this problem, for some reason, some points in the neighborhood of (n,m) are weighted more heavily than other points. The reason might be that in the past this weighting gave some type of suboptimal performance. In any case, from a parallel signal processing point of view, we have

$$g = f * h$$

where

$$h = \begin{pmatrix} 2/10 & 1/10 \\ \boxed{4/10} & 3/10 \end{pmatrix}^{0}$$

Indeed, using the parallel formulation

$$g = ADD_{(n,m)\in \text{coz}(h)}[SCALAR(TRAN(f;n,m);h(n,m)]$$

illustrated in block diagram form in Figure 3.7, gives

$$g = \frac{2}{10}f_{0,1} + \frac{4}{10}f_{0,0} + \frac{1}{10}f_{1,1} + \frac{3}{10}f_{1,0}$$

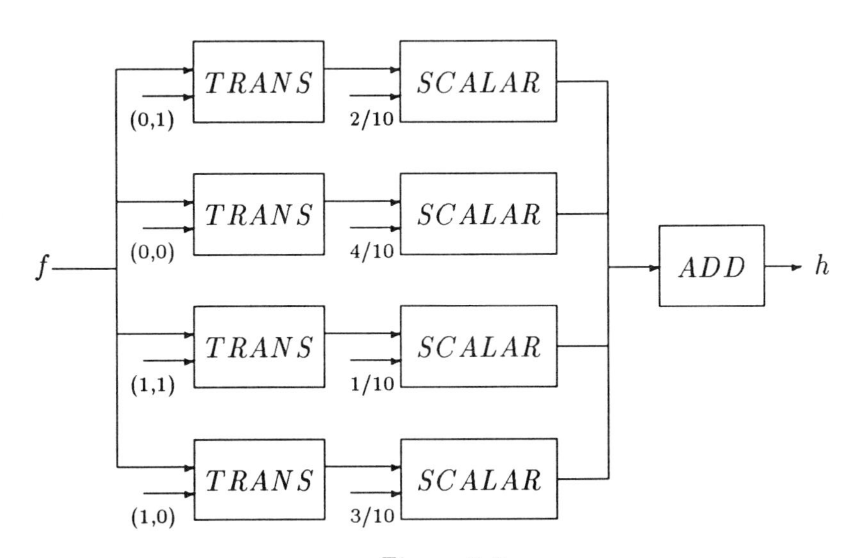

Figure 3.7
Moving Average Parallel Convolver

To validate the parallel formulation, notice that

$$f(r,s)_{pq} = f(r-p, s-q)$$

Hence

$$g(n,m) = \frac{2}{10}f(n,m-1) + \frac{4}{10}f(n,m) + \frac{1}{10}f(n-1,m-1) + \frac{3}{10}f(n-1.m)$$

which is the pointwise description. An instance of moving average parallel convolution is given in the formula provided in Figure 3.8 using

$$f = \begin{pmatrix} 2 & 3 \\ 2 & 2 \end{pmatrix}^{0}_{0,0}$$

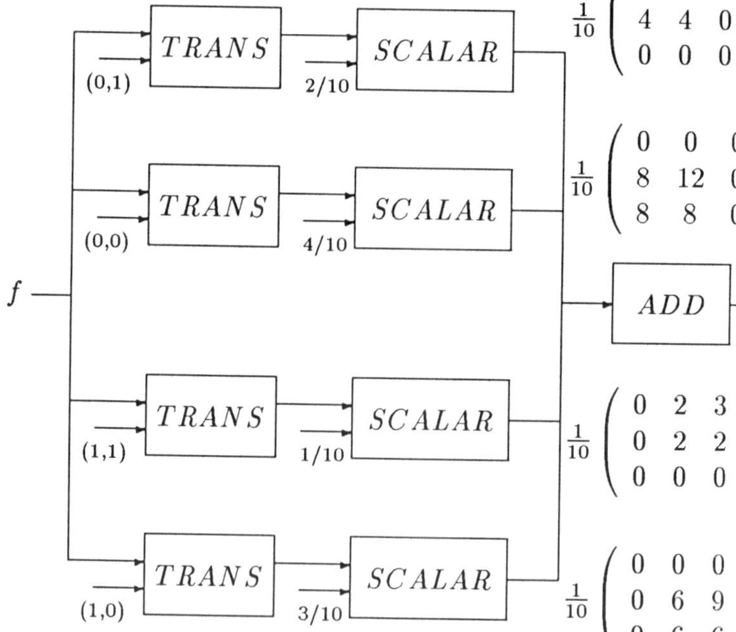

TRANS (0,1) → SCALAR 2/10 → $\frac{1}{10}\begin{pmatrix} 4 & 6 & 0 \\ 4 & 4 & 0 \\ 0 & 0 & 0 \end{pmatrix}^{0}_{0,1}$

TRANS (0,0) → SCALAR 4/10 → $\frac{1}{10}\begin{pmatrix} 0 & 0 & 0 \\ 8 & 12 & 0 \\ 8 & 8 & 0 \end{pmatrix}^{0}_{0,1}$

f → ADD → h

TRANS (1,1) → SCALAR 1/10 → $\frac{1}{10}\begin{pmatrix} 0 & 2 & 3 \\ 0 & 2 & 2 \\ 0 & 0 & 0 \end{pmatrix}^{0}_{0,1}$

TRANS (1,0) → SCALAR 3/10 → $\frac{1}{10}\begin{pmatrix} 0 & 0 & 0 \\ 0 & 6 & 9 \\ 0 & 6 & 6 \end{pmatrix}^{0}_{0,1}$

Figure 3.8

The solution is

$$g = \frac{1}{10}\begin{pmatrix} 4 & 8 & 3 \\ 12 & 24 & 11 \\ 8 & 14 & 6 \end{pmatrix}^{0}_{0,1}$$

Note that we can check the solution obtained in a parallel fashion using the pointwise formula. For instance,

$$g(1,1) = \frac{2}{10}f(1,0) + \frac{4}{10}f(1,1) + \frac{1}{10}f(0,0) + \frac{3}{10}f(0,1) = \frac{4}{5}$$

which is the correct value. Additional points can be checked in this laborious fashion. As an intuitive check, however, notice that the sum of all the values in the signal f is 9. Since g is another signal whose values are obtained by using an average of the values of f, the sum of its values must also be 9.

An important commonly used moving average filter is defined pointwise by

$$g(n,m) =$$

$$\frac{f(n,m) + f(n-1,m) + f(n+1,m) + f(n,m-1) + f(n,m+1)}{5}$$

Here we average all the strong neighbors formed using an equal weighting. The parallel formulation for this filter is

$$g = f * h$$

with

$$h = \frac{1}{5}\begin{pmatrix} 0 & 1 & 0 \\ 1 & \boxed{1} & 1 \\ 0 & 1 & 0 \end{pmatrix}^{0}$$

Thus we have

$$g = \frac{1}{5}f + \frac{1}{5}f_{1,0} + \frac{1}{5}f_{-1,0} + \frac{1}{5}f_{0,1} + \frac{1}{5}f_{0,-1}$$

Figure 3.9 provides a block diagram illustration of this filtering

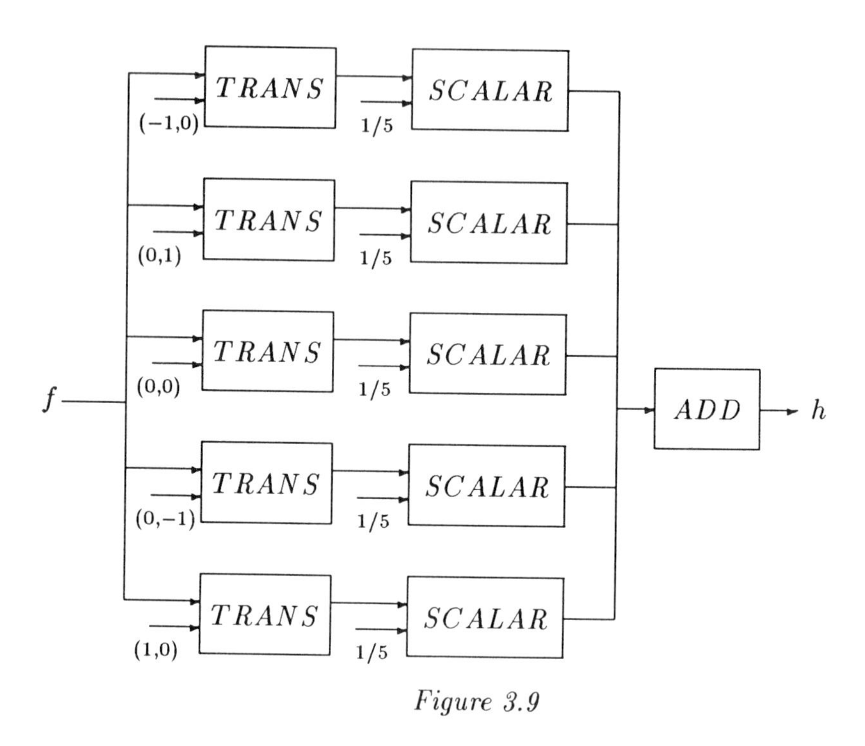

Figure 3.9

Example 3.12 *Apply the moving average filter given in Figure* 3.9 *to the following signal, f.*

$$f = \begin{pmatrix} 2 & 3 \\ 2 & 2 \end{pmatrix}_{0,0}^{0}$$

The result of applying this filter to f is

$$g = f * h = \frac{1}{5} \begin{pmatrix} 0 & 2 & 3 & 0 \\ 2 & 7 & 7 & 3 \\ 2 & 6 & 7 & 2 \\ 0 & 2 & 2 & 0 \end{pmatrix}_{-1,1}^{0}$$

We can check our work several ways. Perhaps the simplest partial check is to notice that the sum of all values in f and g is 9.

Not only can moving average low-pass filtering be implemented in a parallel manner, so also can high-pass filtering. These types of filters enhance rapid variations in signals and attenuate those places of little variation. Again results are described in a pointwise manner, but can be formulated using the unified signal algebra operation presented herein. Additionally, the principle of locality is also frequently used for high-pass filters. Output values are therefore given as a function of input values lying in a neighborhood of the specified output value. Parallel descriptions of such signals have a support region with a neighborhood centered about the origin.

For instance, variation in the horizontal direction is often found by using any of the structuring signals:

$$h = \begin{pmatrix} \boxed{1} & -1 \end{pmatrix}^0$$

$$h = \begin{pmatrix} 1 & -1 \\ \boxed{1} & -1 \end{pmatrix}^0$$

$$h = \begin{pmatrix} 1 & -1 \\ \boxed{1} & -1 \\ 1 & -1 \end{pmatrix}^0$$

Similarly, variation in the vertical direction can be found using any of the following structuring signals:

$$h = \begin{pmatrix} \boxed{1} \\ -1 \end{pmatrix}^0$$

$$h = \begin{pmatrix} \boxed{1} & 1 \\ -1 & -1 \end{pmatrix}^0$$

$$h = \begin{pmatrix} 1 & \boxed{1} & 1 \\ -1 & -1 & -1 \end{pmatrix}^0$$

Variation in the 135° direction is often found using

$$h = \begin{pmatrix} \boxed{0} & 1 \\ -1 & 0 \end{pmatrix}^0$$

and in the 45° direction we can employ

$$h = \begin{pmatrix} 1 & 0 \\ \boxed{0} & -1 \end{pmatrix}^{0}$$

to find the variation. For instance, an illustration of the 45° parallel variation detection formulation is given in Figure 3.10.

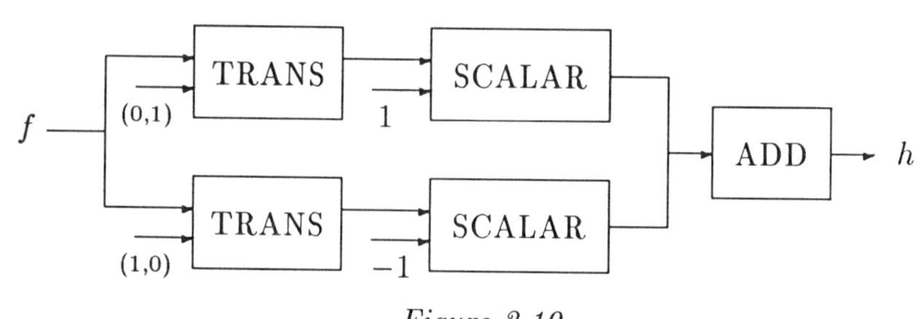

Figure 3.10
Parallel Highpass Filter Convolver

Example 3.13 *Suppose that the digital signal f is given where*

$$f = \begin{cases} 5 & n \geq m \\ 0 & otherwise \end{cases}$$

Accordingly, we can represent f using bound matrices as

$$f = \begin{pmatrix} & & & & \cdot & & & \\ & & & & \cdot & & & \\ & & & & \cdot & & & \\ & 0 & 0 & 0 & 0 & 5 & \\ & 0 & 0 & 0 & 5 & 5 & \\ \cdots & 0 & 0 & \boxed{5} & 5 & 5 & \cdots \\ & 0 & 5 & 5 & 5 & 5 & \\ & 5 & 5 & 5 & 5 & 5 & \\ & & & \cdot & & & \\ & & & \cdot & & & \\ & & & \cdot & & & \end{pmatrix}$$

If f is utilized in the parallel algorithm given in Figure 3.10 for determining 45° variation we obtain output g where

$$g = f_{0,1} - f_{1,0}$$

We have,

$$-f_{1,0} = \begin{pmatrix} & & & \vdots & & & \\ & 0 & 0 & 0 & 0 & -5 & \\ \cdots & 0 & 0 & \boxed{0} & -5 & -5 & \cdots \\ & 0 & 0 & -5 & -5 & -5 & \\ & 0 & -5 & -5 & -5 & -5 & \\ & & & \vdots & & & \end{pmatrix}$$

and

$$f_{0,1} = \begin{pmatrix} & & & \vdots & & & \\ & 0 & 0 & 0 & 5 & 5 & \\ & 0 & 0 & 5 & 5 & 5 & \\ \cdots & 0 & 5 & \boxed{5} & 5 & 5 & \cdots \\ & 5 & 5 & 5 & 5 & 5 & \\ & & & \vdots & & & \end{pmatrix}$$

Thus, we combine the two signals and obtain

$$f * h = \begin{pmatrix} & & & \vdots & & & \\ & 0 & 0 & 0 & 5 & 5 & \\ & 0 & 0 & 5 & 5 & 0 & \\ \cdots & 0 & 5 & \boxed{5} & 0 & 0 & \cdots \\ & 5 & 5 & 0 & 0 & 0 & \\ & & & \vdots & & & \end{pmatrix}$$

The result given in the previous example shows how a high-pass filter can be used in emphasizing regions where variations exist. In general, high-pass filters look for changes in signals and provide responses where these occur. In practice, judicious choices of structuring signals h are made using heuristic rules of thumb, as well as past experience.

3.8 Correlation

An operation somewhat similar to convolution is correlation. This operation is denoted by COR and formally, it is defined at any lattice point (n, m) by

$$COR(f,g)(n,m) = \sum_{k,j=-\infty}^{\infty} f(k,j)g(k-n,j-m)$$

The block diagram illustrating this relationship is

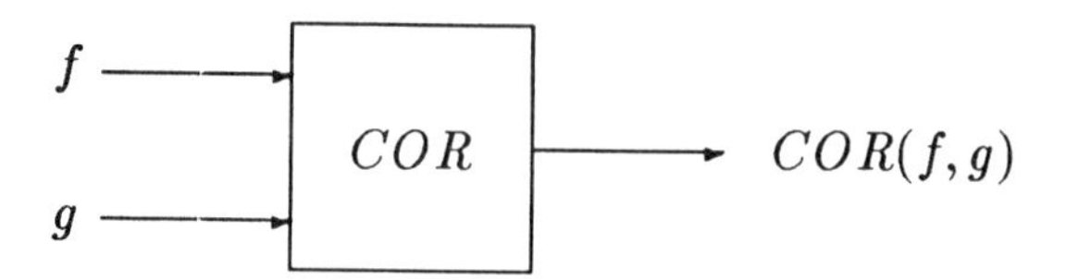

Assuming that the summation converges, and letting $i = k - n$, $r = j - m$, yields an equivalent from of correlation

$$COR(f,g)(n,m) = \sum_{r,i=-\infty}^{\infty} f(i+n,r+m)g(i,r)$$

This implies that

$$COR(f,g)(n,m) = COR(g,f)(-n,-m)$$

or

$$COR(f,g) = NINETY^2[COR(g,f)]$$

and so the operation of correlation is not commutative, in general. However, if $f = g$ great symmetries arise due to the above identity. We will now provide conditions under which the correlation exists.

A sufficient condition for the existence of the correlation function is that f and g are in l_1. Additionally, the correlation function exists if f and g are digital signals where g is of finite support. In this case, COR is given by a finite sum. Of special importance is when both f and g are of finite support. When this is the case we have the minimal computational form for correlation, similar to what was given for convolution. Indeed,

$$COR(f,g)(n,m) =$$

$$\begin{cases} \displaystyle\sum_{\substack{(k,j)\in \mathrm{coz}(f) \\ (k-n,j-m)\in \mathrm{coz}(g)}} f(k,j)g(k-n,j-m) & (n,m) \in D(\mathrm{coz}(f), \mathrm{coz}(N^2(g))) \\ 0 & otherwise \end{cases}$$

Thus, before finding $COR(f,g)$ it is advisable to determine the dilation of $\mathrm{coz}(f)$ with $\mathrm{coz}(N^2(g))$. This dilation set will be a support region of $COR(f,g)$.

Example 3.14 *Let*

$$f = \begin{pmatrix} 2 & 1 \\ -1 & 3 \end{pmatrix}_{0,0}^{0}$$

and

$$g = \begin{pmatrix} 0 & 1 \\ -1 & 3 \end{pmatrix}_{0,0}^{0}$$

We will find the correlation of f with g. First, we must determine a support region for the result using dilation. Thus, we first find

$$g' = N^2(g) = \begin{pmatrix} 3 & -1 \\ 1 & 0 \end{pmatrix}_{-1,1}^{0}$$

and

$$\mathrm{coz}(f) = \{(0,0),(0,-1),(1,0),(1,-1)\}$$

$$\mathrm{coz}(g) = \{(0,-1),(1,0),(1,-1)\}$$

Also,

$$\mathrm{coz}(N^2(g)) = \{(-1,1),(-1,0),(0,1)\}$$

Therefore,

$$D(\mathrm{coz}(f), \mathrm{coz}(N^2(g))) =$$

$$\{(-1,1),(-1,0),(0,1),(0,0),(-1,-1),(0,-1),(1,1),(1,0)\}$$

So $COR(f,g) = 0$ *outside of this set. The value for points inside this set must be found, one at a time, using the minimal computational formula given above. For instance, we start with* $(n,m) = (0,-1)$*. Specifically,*

$$COR(f,g)(0,-1) =$$

$$\sum_{\substack{(k,j)\in\{(0,0),(0,-1),(1,0),(1,-1)\} \\ (k,j+1)\in\{(0,-1),(1,0),(1,-1)\}}} f(k,j)g(k,j+1) = f(1,-1)g(1,0) = 3$$

Similarly, we obtain

$$\begin{aligned} COR(f,g)(-1,1) &= 6 \\ COR(f,g)(-1,0) &= -1 \\ COR(f,g)(0,1) &= 1 \\ COR(f,g)(0,0) &= 11 \\ COR(f,g)(-1,-1) &= -1 \\ COR(f,g)(1,1) &= -1 \\ COR(f,g)(1,0) &= -3 \end{aligned}$$

Therefore

$$COR(f,g) = \begin{pmatrix} 6 & 1 & -1 \\ -1 & 11 & -3 \\ -1 & 3 & 0 \end{pmatrix}_{-1,1}^{0}$$

Bound matrices prove useful in calculating the correlation of two digital signals of finite support. The procedure, which is similar to convolution, is quite simple to implement. However, unlike convolution there is no need to rotate the signal g. The other details are identical. All that need be done is to pad the minimal bound matrix representation of f with zero to the left, right, above and below. The number of rows of zeros above and below is one less than the number of rows in g, and the number of columns to the left and right is one less than the number columns in g. Thus, a new but equivalent bound matrix is used to represent f. Next, the signal g is translated such that it is located "inside" f. These two signals are multiplied and the sum of all entries are found. This provides the value of COR at the location of the translation of g. The bound matrix g is translated over and over again, the product is formed and the sum taken, providing other values of $COR(f, g)$. An example should make this procedure clearer.

Example 3.15 *Let us again find $COR(f, g)$ where*

$$f = \begin{pmatrix} 2 & 1 \\ -1 & 3 \end{pmatrix}_{0,0}^{0}$$

and

$$g = \begin{pmatrix} 0 & 1 \\ -1 & 3 \end{pmatrix}_{0,0}^{0}$$

First we pad the minimal bound matrix of f with one row and one column of zeros on each side. We obtain

$$f = \begin{pmatrix} 0 & 0 & 0 & 0 \\ 0 & 2 & 1 & 0 \\ 0 & -1 & 3 & 0 \\ 0 & 0 & 0 & 0 \end{pmatrix}_{-1,1}^{0}$$

Next we form all translates of g which "fit inside" this representation of f. There are nine such. To begin with, we have

$$g_{-1,1} = \begin{pmatrix} 0 & 1 \\ -1 & 3 \end{pmatrix}_{-1,1}^{0}$$

So $COR(f, g)(-1, 1)$ is found by forming the dot product of $g_{-1,1}$ with

$$\begin{pmatrix} 0 & 0 \\ 0 & 2 \end{pmatrix}_{-1,1}^{0}$$

thereby obtaining 6.

$$g_{-1,0} = \begin{pmatrix} 0 & 1 \\ -1 & 3 \end{pmatrix}_{-1,0}^{0}$$

$COR(f, g)(-1, 0)$ *is found by forming the dot product of* $g_{-1,0}$ *with*

$$\begin{pmatrix} 0 & 2 \\ 0 & -1 \end{pmatrix}_{-1,0}^{0}$$

thereby obtaining -1. *Similarly, we get*

$$\begin{aligned} COR(f, g)(0, 1) &= 1 \\ COR(f, g)(0, 0) &= 11 \\ COR(f, g)(-1, -1) &= -1 \\ COR(f, g)(1, 1) &= -1 \\ COR(f, g)(1, 0) &= -3 \end{aligned}$$

Therefore, we again obtain

$$COR(f, g) = \begin{pmatrix} 6 & 1 & -1 \\ -1 & 11 & -3 \\ -1 & 3 & 0 \end{pmatrix}_{-1,1}^{0}$$

As in the case of convolution, a parallel representation exists for determining the correlation. Indeed, for f and g in $\Re^{Z \times Z}$, with g of finite support, we have

$$COR(f, g) =$$

$$ADD_{(n,m) \in coz(N^2(g))}[SCALAR(TRAN(f; n, m); NINETY^2(g)(n, m)]$$

Figure 3.11 illustrates the block diagram corresponding to this implementation of correlation. The blocks are identical to those used in the convolution expression except that the $NINETY^2$ block is also needed for correlation.

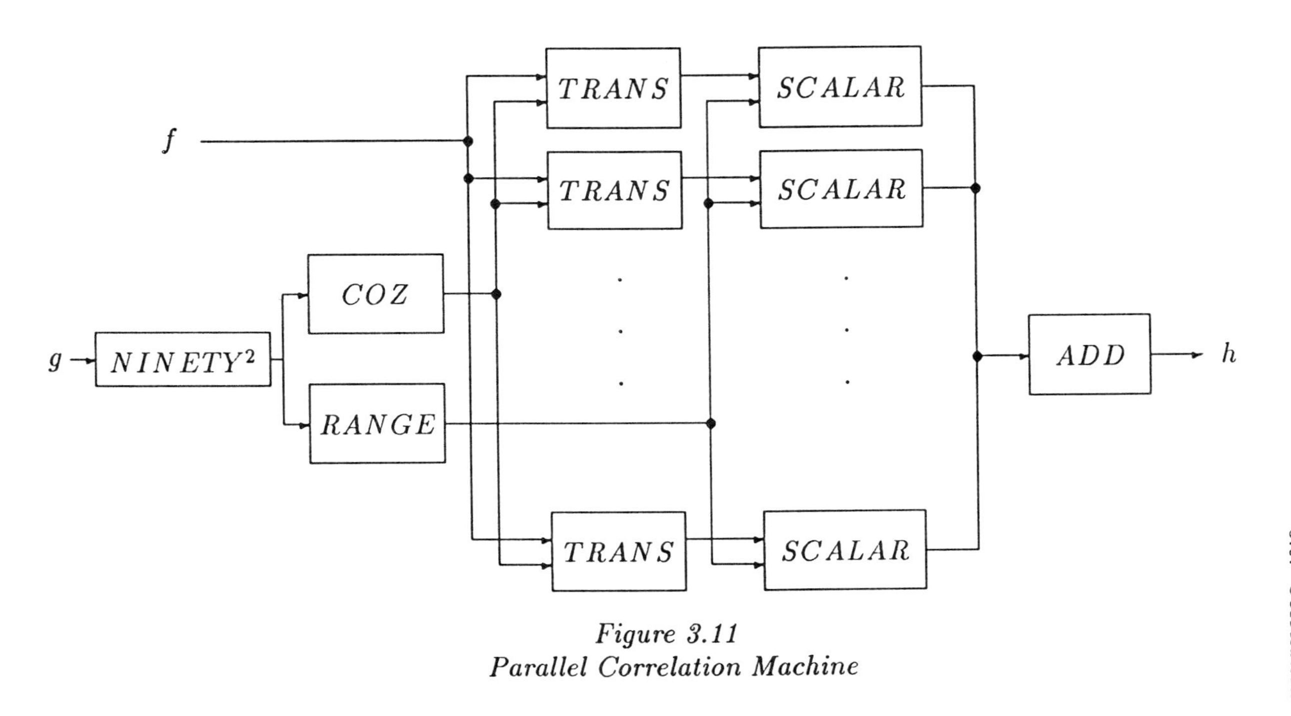

Figure 3.11
Parallel Correlation Machine

Example 3.16 *Now we will use the parallel algorithm for finding the correlation of*

$$f = \begin{pmatrix} 2 & 1 \\ -1 & 3 \end{pmatrix}^{0}_{0,0}$$

and

$$g = \begin{pmatrix} 0 & 1 \\ -1 & 3 \end{pmatrix}^{0}_{0,0}$$

The first operation to be performed is 180° *rotation*

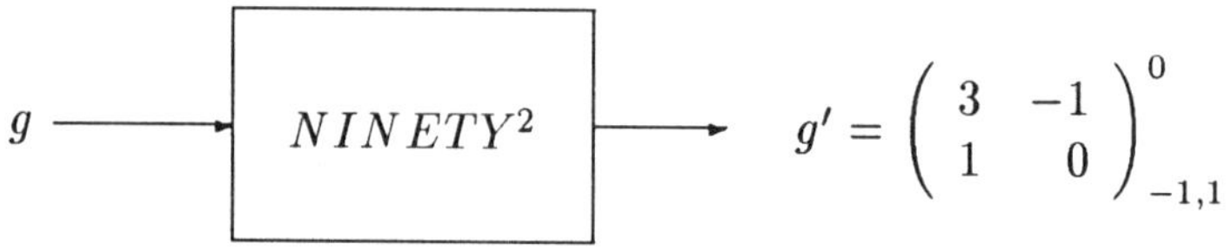

$$g \longrightarrow \boxed{NINETY^2} \longrightarrow g' = \begin{pmatrix} 3 & -1 \\ 1 & 0 \end{pmatrix}^{0}_{-1,1}$$

Next, the co-zero set is found for g' *along with the corresponding real values for each point in the set. These values are stored in the stacks*

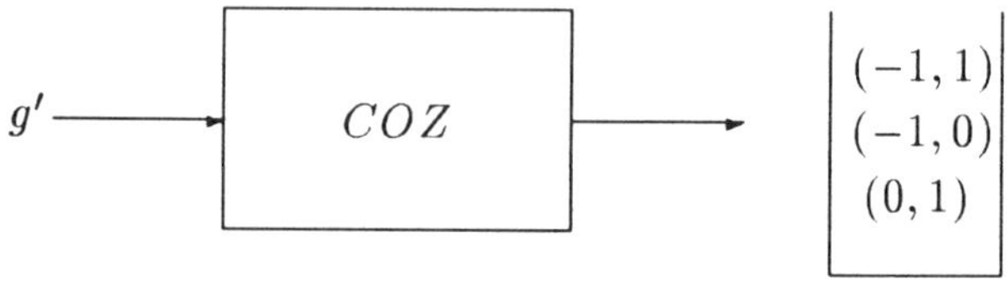

$$g' \longrightarrow \boxed{COZ} \longrightarrow \begin{array}{|c|} (-1,1) \\ (-1,0) \\ (0,1) \\ \hline \end{array}$$

and

$$g' \longrightarrow \boxed{RANGE} \longrightarrow \begin{array}{|c|} 3 \\ 1 \\ -1 \\ \hline \end{array}$$

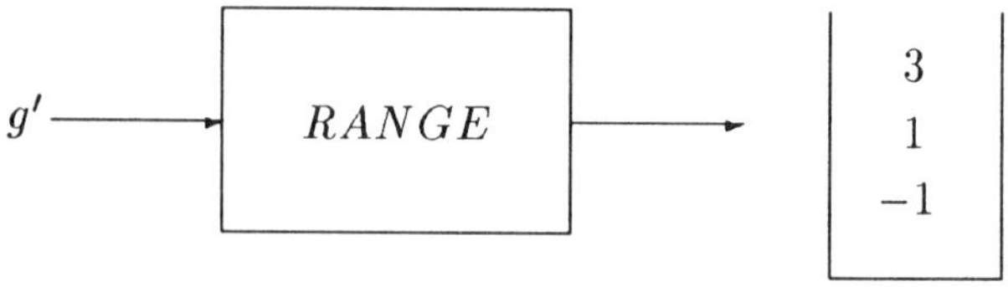

The values in the COZ *stack and* $RANGE$ *stack are sent to the translation block and the scalar block respectively. See Figure* 3.12.

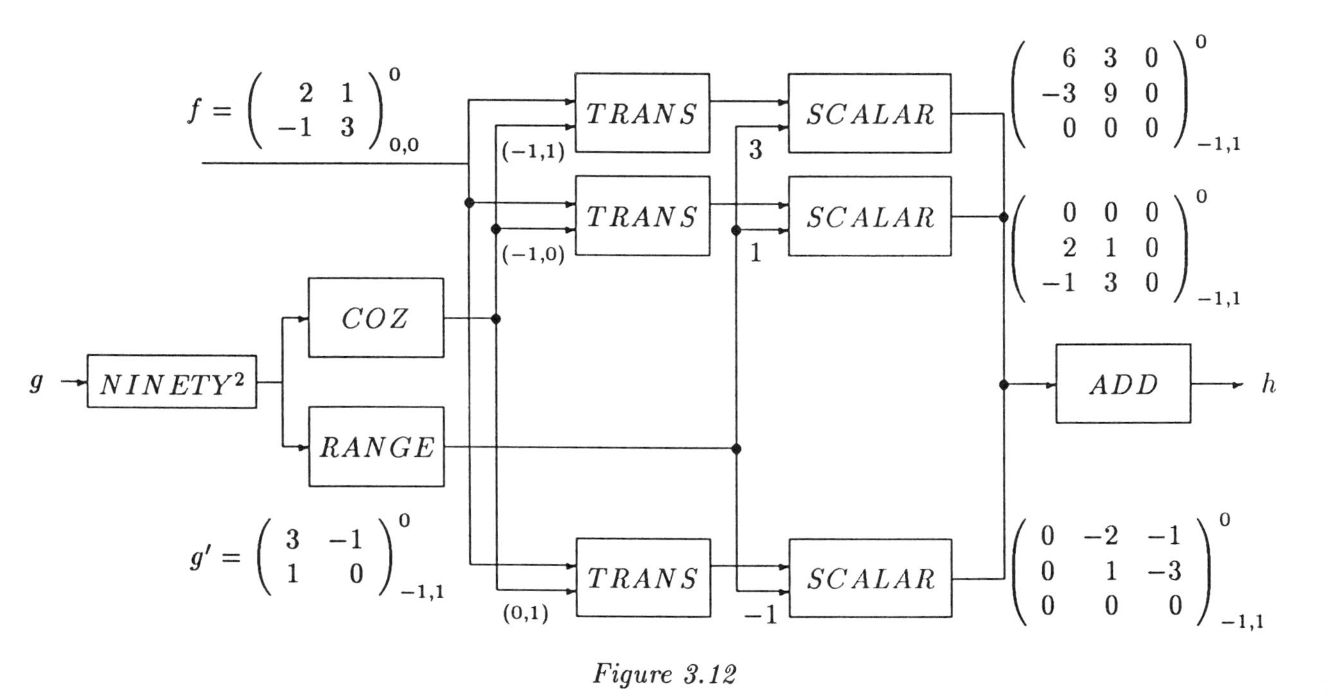

f =
g
NINETY²
COZ
RANGE
g′ =
TRANS
TRANS
TRANS
(−1,1)
(−1,0)
(0,1)
SCALAR
SCALAR
SCALAR
3
1
−1
ADD
h

Figure 3.12

The output of the scalar blocks are added to give the final result

$$COR(f,g) = \begin{pmatrix} 6 & 1 & -1 \\ -1 & 11 & -3 \\ -1 & 3 & 0 \end{pmatrix}_{-1,1}^{0}$$

At the beginning of the section we saw that

$$COR(f,g)(n,m) = COR(g,f)(-n,-m)$$

which means that

$$COR(f,g) = NINETY^2(COR(f,g))$$

In terms of block diagrams - the output of

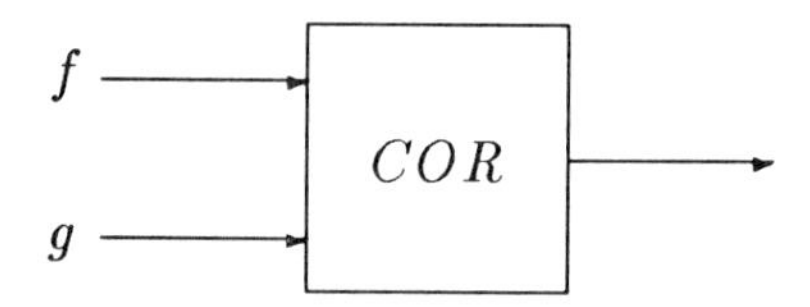

is the same as the output of

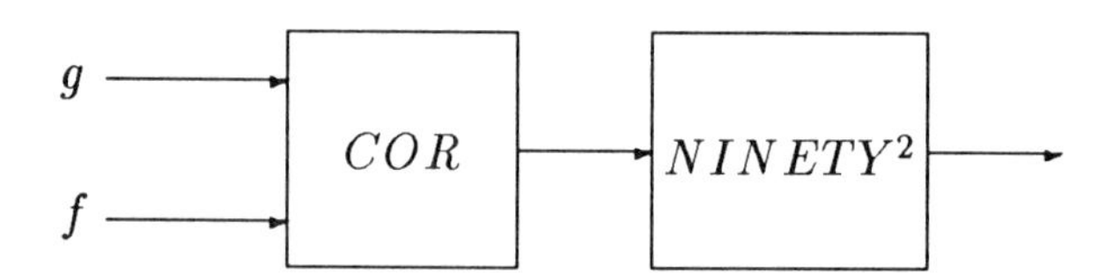

Example 3.17 *Let us find $COR(g,f)$ where*

$$f = \begin{pmatrix} 2 & 1 \\ -1 & 3 \end{pmatrix}_{0,0}^{0}$$

and

$$g = \begin{pmatrix} 0 & 1 \\ -1 & 3 \end{pmatrix}_{0,0}^{0}$$

as in Example 3.16. *This time we find*

$$f' = NINETY^2(f) = \begin{pmatrix} 3 & -1 \\ 1 & 2 \end{pmatrix}^{0}_{-1,1}$$

and then find

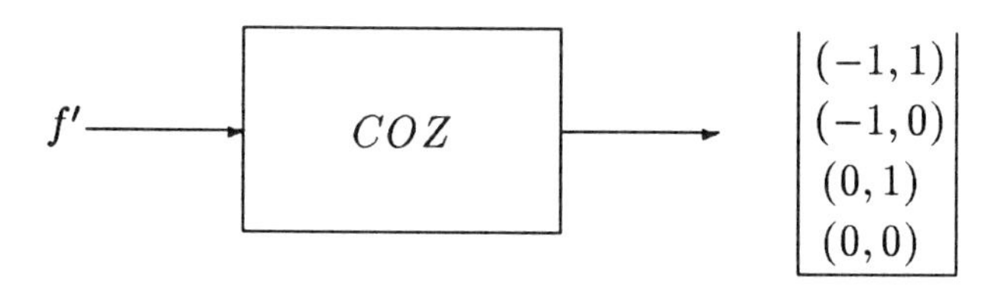

and

$$f' \longrightarrow \boxed{RANGE} \longrightarrow \begin{array}{|c|} 3 \\ 1 \\ -1 \\ 2 \\ \hline \end{array}$$

These values are utilized in the translation and scaling of g. *Accordingly, we obtain*

$$3g_{-1,1} = \begin{pmatrix} 0 & 3 \\ -3 & 9 \end{pmatrix}^{0}_{-1,1}$$

$$g_{-1,0} = \begin{pmatrix} 0 & 1 \\ -1 & 3 \end{pmatrix}^{0}_{-1,0}$$

$$-g_{0,1} = \begin{pmatrix} 0 & -1 \\ 1 & -3 \end{pmatrix}^{0}_{0,1}$$

$$2g_{0,0} = \begin{pmatrix} 0 & 2 \\ -2 & 6 \end{pmatrix}^{0}_{0,0}$$

Adding these bound matrices gives

$$COR(g, f) = \begin{pmatrix} 0 & 3 & -1 \\ -3 & 11 & -1 \\ -1 & 1 & 6 \end{pmatrix}^{0}_{-1,1}$$

The implementation of correlation using the parallel algorithm is left to the reader.

Notice that if we find the 180° rotation of $COR(g,f)$

$$COR(g,f) \longrightarrow \boxed{NINETY^2} \longrightarrow \begin{pmatrix} 6 & 1 & -1 \\ -1 & 11 & -3 \\ -1 & 3 & 0 \end{pmatrix}^{0}_{-1,1}$$

This is the same answer as found in Example 3.17.

The following two identities precisely relate the convolution and the correlation operations. Indeed,

$$COR(f,g) = CONV(f, NINETY^2(g))$$

$$COR(f,g) = NINETY^2(CONV(NINETY^2(f), g)))$$

The second relation holds since

$$COR(g,f) = NINETY^2(COR(f,g))$$

That is, the output of

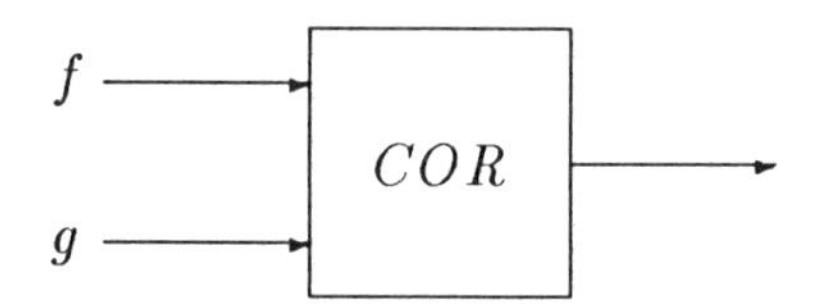

and

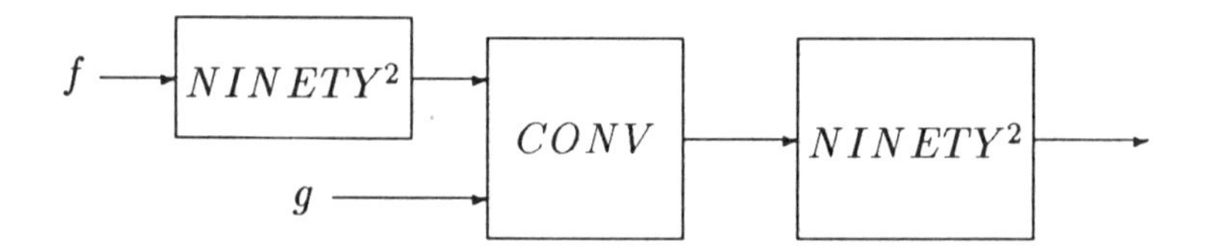

are always identical.

3.9 Applications of Correlation

When forming the correlation of two distinct signals f and g it is customary to call $COR(f,g)$ the cross correlation function. If the two functions f and g are equal, the resulting correlation function, $COR(f,f)$, is called the auto-correlation function. In communications, radar, image processing and other disciplines where pattern recognition plays a major role, the receiving signal is compared using the correlation technique, described in section 3.8, with a structuring signal g. Intuitively, if the auto-correlation function of g is obtained, then the receiving signal f is recognized to be equal to g. If the correlation function obtained is not an auto-correlation function, a different structuring element is used and the correlation operation is performed again. A more in-depth and accurate description of the recognition process using correlation is given next. The essence of the ensuing discussion is that there are simple ways of determining if a correlation is auto-correlation or not.

In order to fully appreciate the application of correlation, the space l_2 will be introduced. We note that both spaces contain all the functions with finite support. The digital signal

$$f = \begin{pmatrix} & \vdots & \vdots & \vdots & \\ \cdots & f(-1,1) & f(0,1) & f(1,1) & \cdots \\ \cdots & f(-1,0) & \boxed{f(0,0)} & f(1,0) & \cdots \\ \cdots & f(-1,-1) & f(0,-1) & f(1,-1) & \cdots \\ & \vdots & \vdots & \vdots & \end{pmatrix}$$

is in l_2 if

$$\sum_{n,m=-\infty}^{\infty} f^2(n,m) < \infty$$

Signals in l_2 are also called finite energy signals. It follows that the sum of two finite energy signal, as well as the scalar multiplication of a finite energy signal is also of finite energy. The value

$$\sqrt{\sum_{n,m=-\infty}^{\infty} f^2(n,m)}$$

is called the two norm of the finite energy signal f and is denoted by $\|f\|_2$.

The digital signal

$$f = \begin{pmatrix} \cdot & \cdot & \cdot & \cdot & \\ \cdot & \cdot & \cdot & \cdot & \\ \cdot & \cdot & \cdot & \cdot & \\ 1/3 & 0 & 0 & 0 & \cdots \\ 1/2 & 0 & 0 & 0 & \cdots \\ 1 & 0 & 0 & 0 & \cdots \\ \boxed{0} & 1 & 1/2 & 1/3 & \cdots \end{pmatrix}$$

is in l_2 since

$$\|f\|_2 = \sqrt{2\sum_{n=1}^{\infty} \frac{1}{n^2}} = \sqrt{\frac{\pi^2}{3}}$$

Notice that f is not in l_1. As was the case for the 1-norm of l_1 functions, the 2-norm for finite energy signals is always a real scalar, that is

$$\|\cdot\|_2 : l_2 \to \Re$$

Moreover it satisfies the three characteristics of a norm.

E1) Positive Definite: $\|f\|_2 \geq 0$ and $\|f\|_2 = 0$ iff $f = 0_{Z\times Z}$

E2) Homogeneous: $\|af\|_2 = |a|\,\|f\|_2$

E3) Triangle Inequality: $\|f+g\|_2 \leq \|f\|_2 + \|g\|_2$

Properties **E1)** and **E2)** can be used in showing an important inequality regarding the correlation. The result is known as the Cauchy-Buniakovski-Schwarz inequality, or CBS inequality.

Let $< f, g >$ represent the summation

$$\sum_{n,m=-\infty}^{\infty} f(n,m)g(n,m)$$

where g is given in a manner similar to f.

CBS Inequality: Suppose that f and g are in l_2, then

$$|< f, g >| \leq \|f\|_2 \cdot \|g\|_2$$

Moreover, the equality sign holds iff f and g are scalar multiples of each other. The result follows by forming

$$\|af + bg\|_2^2 = \sum_{n,m=-\infty}^{\infty} [af(n,m) + bg(n,m)]^2$$

where a and b are some real numbers. By **E1)**

$$a^2 \sum_{n,m=-\infty}^{\infty} f^2(n,m) + \left(2ab \sum_{n,m=-\infty}^{\infty} f(n,m)g(n,m)\right)$$
$$+ \left(b^2 \sum_{n,m=-\infty}^{\infty} g^2(n,m)\right) \geq 0$$

Choose

$$a = - \sum_{n,m=-\infty}^{\infty} f(n,m)g(n,m)$$

and

$$b = \sum_{n,m=-\infty}^{\infty} f^2(n,m)$$

then we obtain

$$\sum_{n,m=-\infty}^{\infty} f^2(n,m) \left[\left(\sum_{n,m=-\infty}^{\infty} f(n,m)g(n,m) \right)^2 \right.$$
$$-2 \left(\sum_{n,m=-\infty}^{\infty} f(n,m)g(n,m) \right)^2$$
$$\left. + \sum_{n,m=-\infty}^{\infty} f^2(n,m) \sum_{n,m=-\infty}^{\infty} g^2(n,m) \right] \geq 0$$

Assume

$$\sum_{n,m=-\infty}^{\infty} f^2(n,m) \neq 0$$

that is

$$\|f\|_2 \neq 0$$

(if $\|f\|_2 = 0$, then $f(n,m) = 0$ for all (n,m) and both sides in the CBS inequality equal zero.) Dividing by

$$\sum_{n,m=-\infty}^{\infty} f^2(n,m)$$

gives

$$-\left[\sum_{n,m=-\infty}^{\infty} f(n,m)g(n,m)\right]^2 + \sum_{n,m=-\infty}^{\infty} f^2(n,m) \sum_{n,m=-\infty}^{\infty} g^2(n,m) \geq 0$$

thus

$$|< f,g >| \leq \|f\|_2 \cdot \|g\|_2$$

The only time equality holds is for

$$\|af + bg\|_2 = 0$$

for some real numbers a and b if

$$af + bg = 0$$

Which means that the two functions f and g are scalar multiples of each other.

The relation between CBS inequality and correlation will be explained next. First, notice that

$$< f, g_{k,j} >= COR(f,g)(k,j)$$

Next, since the norm is found by summing over squares of all values of f, that is

$$\|f\|_2 = \sum_{n,m=-\infty}^{\infty} f^2(n,m)$$

it follows that any translates of f have the same norm as f, that is

$$\|f_{n,m}\|_2 = \|f\|_2$$

By CBS inequality,

$$|COR(f,g)(k,j)| \leq \|f\|_2 \cdot \|g\|_2$$

for all lattice points (k,j). Moreover, we know that equality holds if and only if f is a multiple of $g_{k,j}$. Assuming that f or g are not identically zero, we have

$$\frac{|COR(f,g)(k,j)|}{\|f\|_2 \cdot \|g\|_2} \leq 1$$

The "closer" f and g are, the closer the left side, of the above inequality, is to 1. Herein lies the basis of the correlation technique for the pattern matching of signals.

Example 3.18 *Suppose we are given a database consisting of three signals g, h and r to be used in the recognition process. Let*

$$g = \begin{pmatrix} 1 & 2 \\ 0 & 2 \end{pmatrix}_{0,0}^{0}$$

$$h = \begin{pmatrix} -1 & 2 \\ 0 & 2 \end{pmatrix}_{1,0}^{0}$$

and

$$r = \begin{pmatrix} 3 & 2 & 1 \\ 1 & 1 & 0 \end{pmatrix}_{0,0}^{0}$$

The observed signal is

$$f = \begin{pmatrix} -1 & 2 \\ 0 & 2 \end{pmatrix}_{-3,0}^{0}$$

Correlation techniques will be employed to detect which signal f is "shaped like". More precisely, assume f is a scalar multiple of a translate of one of the signals h, g, or r. We will determine the one f is a scalar multiple of, along with the amount of translation. To solve this problem we form the correlation of f with each of g, h, r and then perform the normalization process. We have

$$COR(g, f) = \begin{pmatrix} 2 & 4 & 0 \\ 2 & 7 & -2 \\ 0 & 4 & -2 \end{pmatrix}_{2,1}^{0}$$

$$COR(h, f) = \begin{pmatrix} 2 & 4 & 0 \\ -2 & 9 & -2 \\ 0 & 4 & -2 \end{pmatrix}_{3,1}^{0}$$

$$COR(r, f) = \begin{pmatrix} 6 & 4 & 2 & 0 \\ 8 & 3 & 0 & -1 \\ 2 & 1 & -1 & 0 \end{pmatrix}_{2,1}^{0}$$

and the norms

$$\|f\|_2 = 3$$
$$\|g\|_2 = 3$$
$$\|h\|_2 = 3$$
$$\|r\|_2 = 4$$

The normalization gives

$$\frac{COR(g, f)}{\|f\|_2 \cdot \|g\|_2} = \frac{1}{9}\begin{pmatrix} 2 & 4 & 0 \\ 2 & 7 & -2 \\ 0 & 4 & -2 \end{pmatrix}_{2,1}^{0}$$

$$\frac{COR(h, f)}{\|f\|_2 \cdot \|h\|_2} = \frac{1}{9}\begin{pmatrix} -2 & 4 & 0 \\ -2 & 9 & -2 \\ 0 & 4 & -2 \end{pmatrix}_{2,1}^{0}$$

$$\frac{COR(r, f)}{\|f\|_2 \cdot \|r\|_2} = \frac{1}{12}\begin{pmatrix} 6 & 4 & 2 & 0 \\ 8 & 3 & 0 & -1 \\ 2 & 1 & -1 & 0 \end{pmatrix}_{2,1}^{0}$$

Notice that the only time the normalized correlation function is of absolute value one is for

$$\frac{COR(h,f)(3,0)}{\|f\|_2 \cdot \|h\|_2}$$

This means that h is a scalar multiple of $f_{3,0}$. Moreover, we known that in this case the scalar multiple is one.

Notice that if an auto-correlation is performed then the correlation function will be largest at the origin, that is

$$COR(f,f)(0,0)$$

will be maximum. Moreover,

$$COR(f,f)(n,m) = COR(f,f)(-n,-m)$$

holds true in this case.

3.10 Exercises

1. Let

$$f = \begin{pmatrix} 2 & 3 & 1 \\ 0 & -1 & 1 \end{pmatrix}_{5,2}^{0}$$

   Show that

$$\text{coz}(f) = \{(5,2),(6,2),(6,1),(7,2),(7,1)\}.$$

   Must $\text{supp}(f) = \text{coz}(f)$? Explain.

2. Find $\text{coz}(g)$ where

$$g = \begin{pmatrix} 4 & 1 \\ -1 & 0 \end{pmatrix}_{0,0}^{0}$$

3. Refer to f and g given in the previous two exercises. Find $D(\text{coz}(f), \text{coz}(g))$.

4. For f and g given in Exercises 1 and 2 respectively, show that

$$f * g = \begin{pmatrix} 8 & 14 & 7 & 1 \\ -2 & -7 & 2 & 1 \\ 0 & 1 & -1 & 0 \end{pmatrix}_{5,2}^{0}$$

   Do this using the definition of convolution.

5. Give an example showing that $\text{coz}(f*g)$ need not equal
$$D(\text{coz}(f), \text{coz}(g)).$$

6. Use the bound matrix method given in section 3.4 to obtain $f * g$ given in Exercise 4.

7. Use the parallel convolution algorithm given in section 3.4 to obtain $f * g$ given in Exercise 4.

8. Find all digital signals g which have the property that $f * g = f$ for all f.

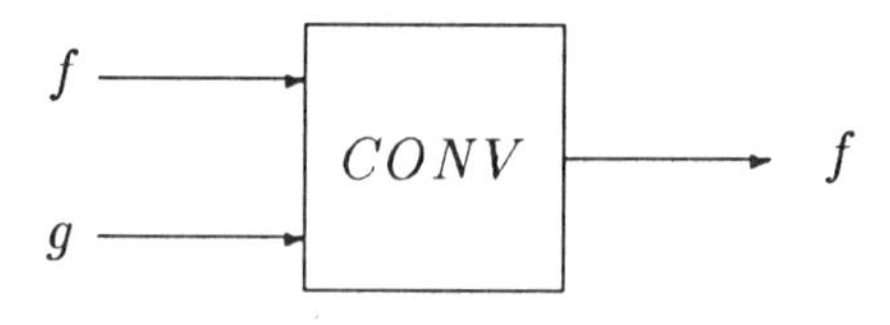

9. Refer to the parallel convolution block diagram given earlier and let $g = \delta_{1,0} = (1)_{1,0}^{0}$, then it is obvious that the output of

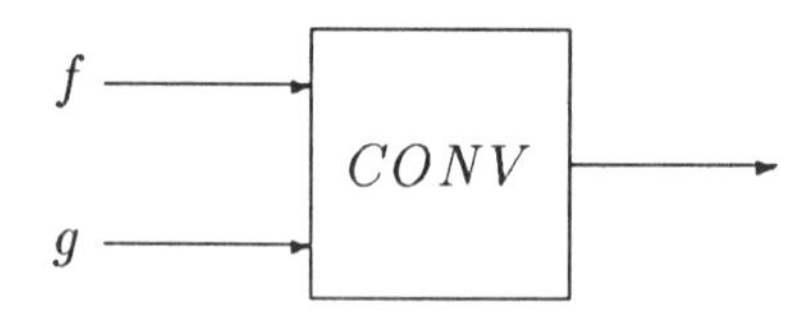

and

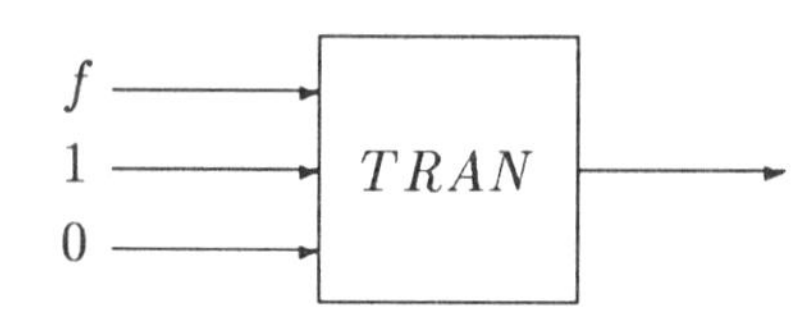

are identical. Generalize the above observation for arbitrary translations.

10. Can we form $f * g$ for any digital signals f and g? Explain

11. What happens if we try to perform $u * u$?

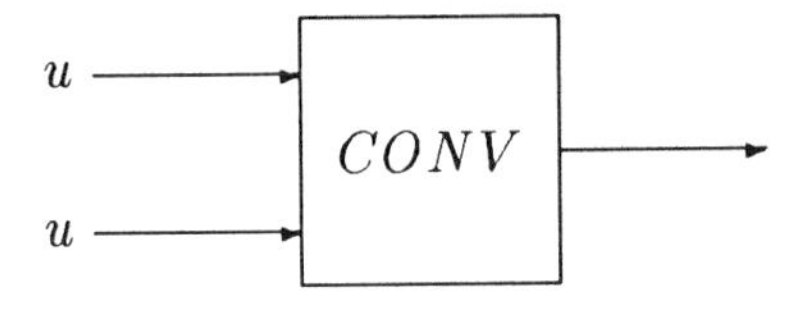

12. Provide a block diagram illustrating the Banach algebra property **B2)** of the convolution given in section 3.6.

13. Use the parallel convolution algorithm to verify Banach algebra property **B5)** for scalar $a = -1$ and

$$f = \begin{pmatrix} 3 & 2 & 1 \end{pmatrix}_{0,0}^{0} \quad \text{and} \quad g = \begin{pmatrix} 1 & 0 \\ 2 & -1 \end{pmatrix}_{0,0}^{0}$$

14. For the signal given in Exercises 1 and 2, illustrate the normed vector space properties **N1)**, **N2)** and **N3)** given in section 3.6.

15. For $f * g$ given in Exercise 4 illustrate the Banach algebra inequality

$$\|f * g\|_1 \leq \|f\|_1 \cdot \|g\|_1$$

16. Show how to perform the weighted moving average

$$g(n) = \frac{3f(n-1,m) + 2f(n,m)}{5}$$

by employing the parallel convolution algorithm. Do this for

$$f = \begin{pmatrix} 1 & 1 & 1 & 1 & 1 & 1 & 1 & 1 \\ 2 & 2 & 2 & 2 & 3 & 3 & 3 & 3 \end{pmatrix}_{0,0}^{0}$$

17. Implement the high-pass filter

$$h = \begin{pmatrix} -1 & 1 \end{pmatrix}^{0}_{0,0}$$

on the signal f given in Exercise 16.

18. Several important structuring signals h are used in finite impulse response filtering (FIR), that is in convolving a signal f with a finite support signal h. These are defined in the horizontal direction. For each structuring h (or windowing signal as they are often called) set up the parallel convolution algorithm to perform the FIR filtering of the first quadrant step signal u. Do this for $N = 7$. In each case, find the range and co-zero stack values, and describe what these FIR filters do in the horizontal as well as in the vertical direction. How can we make a FIR which acts in the vertical as well as in the horizontal direction, similar to the filter given here?

(a) Rectangular window

$$h(k,j) = \begin{cases} 1 & |k| \leq \frac{N-1}{2}, j = 0, N \geq 1 \\ 0 & otherwise \end{cases}$$

(b) Triangular window

$$h(k,j) = \begin{cases} 1 - \frac{2|k|}{N-1} & |k| \leq \frac{N-1}{2}, j = 0, N \geq 2 \\ 0 & otherwise \end{cases}$$

(c) Raised-Cosine window

$$h(k,j) = \begin{cases} \frac{1}{2}[1 + \frac{\cos(2k)}{N-1}] & |k| \leq \frac{N-1}{2}, j = 0, N \geq 2 \\ 0 & otherwise \end{cases}$$

(d) Hamming window

$$h(k,j) = \begin{cases} 0.54 + 0.46 * \cos \frac{2k}{N-1} & |k| \leq \frac{N-1}{2}, \\ & j = 0, \ N \geq 2 \\ 0 & otherwise \end{cases}$$

(e) Blackman window

$h(k,j) =$

$$\begin{cases} 0.42 + 0.5 * \cos\frac{2k}{N-1} + 0.08 * \cos\frac{4k}{N-1} & |k| \leq \frac{N-1}{2}, \\ & j = 0,\ N \geq 2 \\ 0 & otherwise \end{cases}$$

19. Let

$$f = \begin{pmatrix} 2 & 4 \\ 1 & 3 \end{pmatrix}^{0}_{5,0} \quad \text{and } g = \begin{pmatrix} 1 & -1 \\ 0 & 3 \end{pmatrix}^{0}_{2,0}$$

Find $COR(f,g)$ using

(a) the pointwise method.

(b) the bound matrix procedure.

(c) the parallel algorithm of Figure 3.12.

20. For f and g given in exercise 19 illustrate the equational identity given in terms of the block diagram: the output of

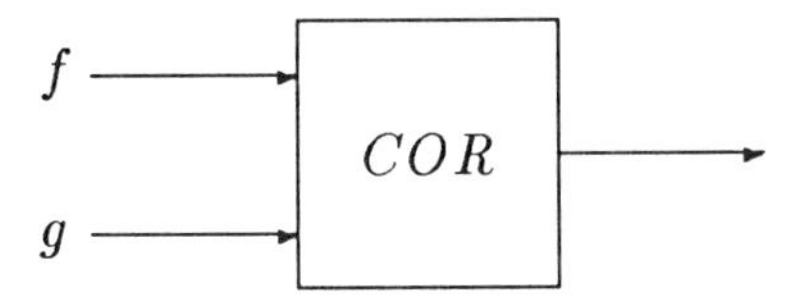

is the same as the output of

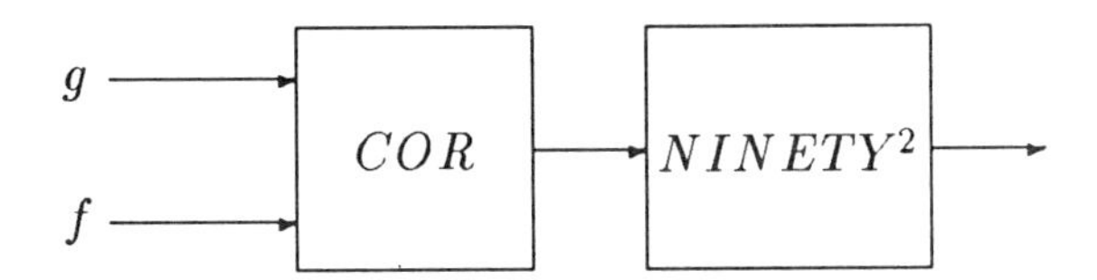

21. Use the signals

$$f = \begin{pmatrix} 2 & 0 & 0 \\ 0 & 1 & 0 \\ 0 & 0 & 3 \end{pmatrix}^{0}_{-1,1} \quad \text{and} \quad g = \begin{pmatrix} 0 & 0 & 2 \\ 0 & 3 & 0 \\ 2 & 0 & 0 \end{pmatrix}^{0}_{-1,1}$$

to find $COR(f,g)$. What did you observe in doing this? Hint: here $N^2(g) = g$.

22. Consider the following three signals

$$f = \begin{pmatrix} 1 & 2 & -1 \\ 0 & 1 & 0 \end{pmatrix}_{0,0}^{0}$$

$$g = \begin{pmatrix} 1 & 0 & -1 \\ 0 & 1 & 0 \end{pmatrix}_{0,0}^{0}$$

and

$$h = \begin{pmatrix} 2 & 0 & -2 \\ 0 & 2 & 0 \end{pmatrix}_{0,0}^{0}$$

Find the auto-correlation and cross-correlation functions involving these signals.

23. Illustrate the CBS inequality for the correlation function involving each of the signals given in Exercise 22.

24. Explain how the properties of the normalized correlation function are used in the pattern recognition application of section 3.9.

25. Which of the digital signals f below are in l_2?

(a) $f(n,m) = \begin{cases} 1/n & n \neq 0, \text{ all } m \\ 0 & n = 0, \text{ all } m \end{cases}$

(b) $f(n,m) = 2^{-|n|-|m|}$

(c) $f(n,m) = 1$

26. Which of the signals in Exercise 25 are in l_1?

27. Show that if $f \in l_1$, then $f \in l_2$. Is the converse true?

4

$\mathcal{Z}$ Transforms

4.1 Formal Introduction to $\mathcal{Z}$ Transforms

One of the most important, and perhaps the best known transform technique, involving digital signals f, in $\Re^Z$, is the $\mathcal{Z}$ transform. A similar transform exists in two dimensions and is also called the $\mathcal{Z}$ transform. It is defined, in a formal manner, by associating the two dimensional infinite series

$$\sum_{n,m=-\infty}^{\infty} a_{n,m} z^{-n} w^{-m}$$

with the two dimensional digital signal

$$\begin{pmatrix} & \cdot & \cdot & \cdot & \\ & \cdot & \cdot & \cdot & \\ & \cdot & \cdot & \cdot & \\ \cdots & a_{-1,1} & a_{0,1} & a_{1,1} & \cdots \\ \cdots & a_{-1,0} & \boxed{a_{0,0}} & a_{1,0} & \cdots \\ \cdots & a_{-1,-1} & a_{0,-1} & a_{1,-1} & \cdots \\ & \cdot & \cdot & \cdot & \\ & \cdot & \cdot & \cdot & \\ & \cdot & \cdot & \cdot & \end{pmatrix}$$

This correspondence is also often denoted by $F(z, w)$,

$$f \longleftrightarrow \sum_{n,m=-\infty}^{\infty} a_{n,m} z^{-n} w^{-m}$$

or

$$\mathcal{Z}(f) = \sum_{n,m=-\infty}^{\infty} a_{n,m} z^{-n} w^{-m}$$

The ordering of this series is unimportant since it is only a formal sum.

In this section, z and w are indeterminate; this means that z and w are just symbols and should not be treated as numbers. Accordingly, no mention of convergence is given to the infinite series. Moreover, z^0 will be identified with the "symbol" number 1, and the same is true for w^0. If the coefficient of $z^k w^j$ is zero, then this term can be left out.

Example 4.1 *If*

$$f = \begin{pmatrix} \cdots & 1 & 1 & 1 & 1 & 1 & 2 & 2^2 & 2^3 & 2^4 & \cdots \\ \cdots & 4 & -3 & 2 & -1 & \boxed{1} & 1 & 2! & 3! & 4! & \cdots \\ \cdots & 0 & 0 & 0 & 0 & 0 & 1 & 2 & 3 & 4 & \cdots \end{pmatrix}^0$$

then

$$\mathcal{Z}(f) = \sum_{n=-\infty}^{-1} (-1)^{n+1} n z^{-n} + \sum_{n=0}^{\infty} n! z^{-n}$$

$$+ \left[\sum_{n=-\infty}^{0} z^{-n} w^{-1} + \sum_{n=1}^{\infty} 2^n z^{-n} w^{-1} \right]$$

$$+ \left[\sum_{n=1}^{\infty} n z^{-n} w \right]$$

Example 4.2 *Suppose that*

$$f = \begin{pmatrix} \cdot & \cdot & \cdot & \\ \cdot & \cdot & \cdot & \\ \cdot & \cdot & \cdot & \\ 4 & 0 & 0 & \cdots \\ 3 & 0 & 0 & \cdots \\ 2 & 0 & 0 & \cdots \\ \boxed{1} & 2^{-1} & 2^{-2} & \cdots \end{pmatrix}^0$$

then

$$\mathcal{Z}(f) = \sum_{n=0}^{\infty} 2^{-n} z^{-n} + \sum_{m=1}^{\infty} (m+1) w^{-m}$$

Example 4.3 *If*

$$f = \begin{pmatrix} \cdot & \cdot & \cdot & \cdot & \\ \cdot & \cdot & \cdot & \cdot & \\ \cdot & \cdot & \cdot & 1/3! & \cdots \\ 0 & 0 & 1/2! & 0 & \cdots \\ 0 & 1/1 & 0 & 0 & \cdots \\ \boxed{1} & 0 & 0 & 0 & \cdots \end{pmatrix}^{0}$$

then

$$\mathcal{Z}(f) = \sum_{n=0}^{\infty} \frac{1}{n!} z^{-n} w^{-n}$$

The set of all infinite series involving integer powers of z and w with real coefficients will be denoted by $\Re[z,w]$. Notice that for every signal f in $\Re^{Z\times Z}$, there will be one and only one infinite series in $\Re[z,w]$ corresponding to it, which of course is the $\mathcal{Z}$ transform. In the opposite direction, for every infinite series in $\Re[z,w]$ there will be a digital signal (in $\Re^{Z\times Z}$) which is its correspondent. Thus, the mapping:

$$\mathcal{Z} : \Re^{Z\times Z} \to \Re[z,w]$$

is one to one and onto. Accordingly we can define the inverse mapping $\mathcal{Z}^{-1}$ where

$$\mathcal{Z}^{-1} : \Re[z,w] \to \Re^{Z\times Z}$$

and

$$\mathcal{Z}^{-1}\left(\sum_{n,m=-\infty}^{\infty} a_{n,m} z^{-n} w^{-m} \right)$$

$$= \begin{pmatrix} & \cdot & \cdot & \cdot & \\ & \cdot & \cdot & \cdot & \\ & \cdot & \cdot & \cdot & \\ \cdots & a_{-1,1} & a_{0,1} & a_{1,1} & \cdots \\ \cdots & a_{-1,0} & a_{0,0} & a_{1,0} & \cdots \\ \cdots & a_{-1,-1} & a_{0,-1} & a_{1,-1} & \cdots \\ & \cdot & \cdot & \cdot & \\ & \cdot & \cdot & \cdot & \\ & \cdot & \cdot & \cdot & \end{pmatrix}$$

In general, notice that

$$\mathcal{Z}^{-1}(\mathcal{Z}(f)) = f$$

If

$$f \longleftrightarrow F(z,w)$$

then

$$\mathcal{Z}(\mathcal{Z}^{-1}(F(z,w))) = F(z,w).$$

Example 4.4 *Notice that the first quadrant step function u has the unique $\mathcal{Z}$ transform*

$$\mathcal{Z}(u) = \sum_{n,m=0}^{\infty} z^{-n} w^{-m}$$

Several additional special $\mathcal{Z}$ transforms should be noted, similar to the one given in Example 4.4. Indeed, we have

$$0_{Z\times Z} \longleftrightarrow 0$$

$$1_{Z\times Z} \longleftrightarrow \sum_{n,m=-\infty}^{\infty} z^{-n} w^{-m}$$

$$\delta \longleftrightarrow 1$$

4.2 Some Operations Involving $\mathcal{Z}$ Transforms

In this section, we define four simple operations for manipulating infinite series in $\Re[z,w]$. These operations are then one by one related to operations involving digital signals.

The first operation given here is addition. Two infinite series, $F(z,w)$ and $G(z,w)$ in $\Re[z,w]$ are added together to yield another series in $\Re[z,w]$. The result is denoted as

$$(F+G)(z,w)$$

or by "The sloppy, but more common, notation"

$$F(z,w) + G(z,w)$$

It is formed by adding the coefficients of equal powers of z and w. That is, if

$$F(z,w) = \sum_{n,m=-\infty}^{\infty} a_{n,m} z^{-n} w^{-m}$$

and

$$G(z,w) = \sum_{n,m=-\infty}^{\infty} b_{n,m} z^{-n} w^{-m}$$

then

$$F(z,w) + G(z,w) = \sum_{n,m=-\infty}^{\infty} (a_{n,m} + b_{n,m}) z^{-n} w^{-m}$$

The second operation involving series in $\Re[z,w]$ is scalar multiplication. Here, any infinite series $F(z,w)$ can be multiplied by a real number a which again results in an infinite series in $\Re[z,w]$, denoted by $(a \cdot F)(z,w)$, or just by $aF(z,w)$. We define

$$aF(z,w) = \sum_{n,m=-\infty}^{\infty} (a \cdot a_{n,m}) z^{-n} w^{-m}$$

The third operation defined on infinite series is multiplication. It, however, involves the multiplication of two elements of $\Re[z,w]$. One of them is just $z^k w^j$ with k and j integers. The other series is arbitrary, say $F(z,w)$. The result is denoted by $(z^k w^j \cdot F)(z)$, or just $z^k w^j F(z)$ and is defined as

$$z^k w^j F(z,w) = \sum_{n,m=-\infty}^{\infty} a_{n,m} z^{k-n} w^{j-m}$$

In particular, notice that

$$z^0 w^0 F(z,w) = F(z,w)$$

and

$$z^{-1} F(z,w) = \sum_{n,m=-\infty}^{\infty} a_{n,m} z^{-1-n} w^{-m}$$

The fourth, and final operation on series in $\Re[z,w]$ is an inversion type operation, denoted by $\mathcal{I}$, where we replace both z by $1/z$ and

w by $1/w$. So

$$(\mathcal{I}(F))(z,w) = \mathcal{I}(F(z,w)) = \sum_{n,m=-\infty}^{\infty} a_{n,m} z^n w^m$$

which we will often write as

$$\sum_{n,m=-\infty}^{\infty} a_{-n,-m} z^{-n} w^{-m}$$

Clearly,

$$\mathcal{I}(\mathcal{I}(F(z,w))) = F(z,w)$$

Henceforth, we will often denote $\mathcal{I}(F(z,w))$ by the equivalent formulation $F(1/z, 1/w)$.

Example 4.5 *Let*

$$F(z,w) = \sum_{n=-\infty}^{\infty} 2^n z^{-n} + \sum_{m=1}^{\infty} w^{-m} z^1$$

and

$$G(z) = \sum_{n=-\infty}^{0} -2^n z^{-n} + \sum_{n=1}^{\infty} 3^n z^{-n} + \sum_{m=1}^{\infty} w^{-m} z^1$$

then

$$F(z,w) + G(z,w) = \sum_{n=1}^{\infty} (3^n + 2^n) z^{-n} + \sum_{m=1}^{\infty} 2w^{-m} z^1$$

$$2F(z,w) = \sum_{n=-\infty}^{\infty} 2^{n+1} z^{-n} + \sum_{m=1}^{\infty} 2w^{-m} z^1$$

$$z^{-3} w^{-2} F(z,w) = \sum_{n=-\infty}^{\infty} 2^n z^{-n-3} w^{-2} + \sum_{m=1}^{\infty} w^{-m-2} z^{-2}$$

and

$$F(1/z, 1/w) = \sum_{n=-\infty}^{\infty} 2^{-n} z^{-n} + \sum_{m=1}^{\infty} w^{-m} z^{-1}$$

Each of the four operations just defined in $\Re[z,w]$ have analogous operations existing in $\Re^{Z\times Z}$. Corresponding operations can be found using the mapping $\mathcal{Z}$ (or $\mathcal{Z}^{-1}$) since the $\mathcal{Z}$ transform is a one to one correspondence between signals in $\Re^{Z\times Z}$ and series in $\Re[z,w]$. Additionally, if

$$f \longleftrightarrow F(z,w)$$

and

$$g \longleftrightarrow G(z,w)$$

then the following properties hold true:

Z1) Additive Property: $\mathcal{Z}(f+g) = F(z,w) + G(z,w)$
Z2) Homogeneous Property: $\mathcal{Z}(af) = aF(z,w)$
Z3) Shifting Property: $\mathcal{Z}(T(f;n,m)) = z^{-n}w^{-m}F(z,w)$
Z4) Rotation Property: $\mathcal{Z}(N^2(f)) = \mathcal{I}(F(z,w)) = F(1/z,1/w)$

These properties show that $\Re^{Z\times Z}$ and $\Re[z,w]$ are isomorphic. Properties **Z1)** and **Z2)** are often combined as the following to denote the linearity of the $\mathcal{Z}$ transform

$$\mathcal{Z}(af+g) = aF(z,w) + G(z,w)$$

Of special interest about **Z3)** is with $n = 1$, $m = 0$. Here, the $\mathcal{Z}$ transform of a shift of f to the right by one unit is $z^{-1}F(z,w)$. A shift of f one unit up has a $\mathcal{Z}$ transform $w^{-1}F(z,w)$. Special symbolism is often employed using block diagrams. Indeed, instead of using the block diagram

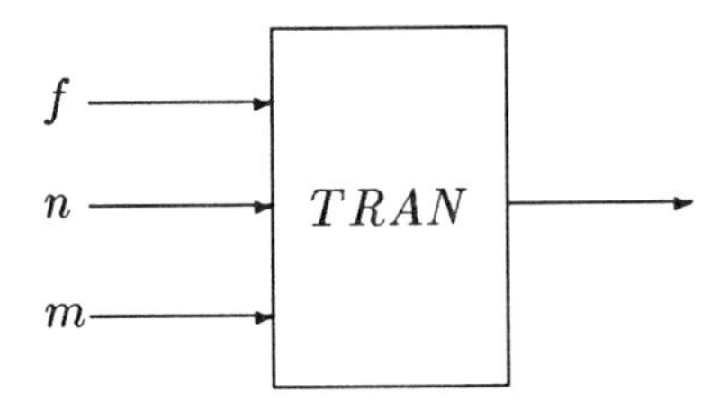

to obtain $TRAN(f;n,m)$, the following block diagram can be employed to specify the same operation.

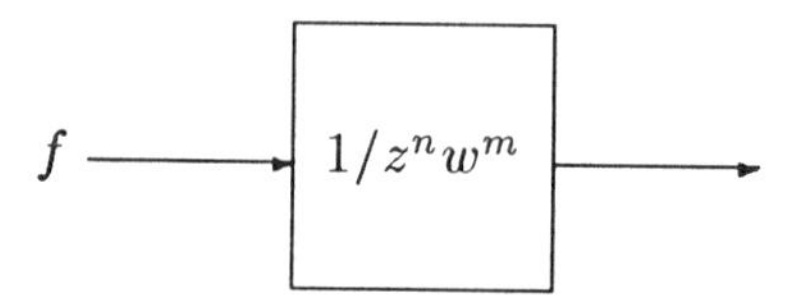

Finally, **Z4)** says that the $\mathcal{Z}$ transform of the 180° rotation of f is found by taking the $\mathcal{Z}$ transform of f itself and then replacing z by $1/z$, w by $1/w$.

Example 4.6 *Suppose that $F(z,w)$ and $G(z,w)$ from Example 4.5 are given, then we have*

$$f \longleftrightarrow F(z,w)$$

and

$$g \longleftrightarrow G(z,w)$$

where

$$f = \begin{pmatrix} & \vdots & \vdots & \vdots & \vdots & \vdots & \\ \cdots & 0 & 1 & 0 & 0 & 0 & \cdots \\ \cdots & 0 & 1 & 0 & 0 & 0 & \cdots \\ \cdots & 2^{-2} & 2^{-1} & \boxed{1} & 2 & 2^2 & \cdots \end{pmatrix}^0$$

and

$$g = \begin{pmatrix} & \vdots & \vdots & \vdots & \vdots & \vdots & \\ \cdots & 0 & 1 & 0 & 0 & 0 & \cdots \\ \cdots & 0 & 1 & 0 & 0 & 0 & \cdots \\ \cdots & -2^{-2} & -2^{-1} & \boxed{-1} & 3 & 3^2 & \cdots \end{pmatrix}^0$$

Moreover,

$$f+g = \begin{pmatrix} \vdots & \vdots & \vdots & \vdots & \\ 2 & 0 & 0 & 0 & \cdots \\ 2 & 0 & 0 & 0 & \cdots \\ 0 & \boxed{0} & 2+3 & 2^2+3^2 & \cdots \end{pmatrix}^0 \longleftrightarrow F(z,w)+G(z,w)$$

$$2f = \begin{pmatrix} & \cdot & \cdot & \cdot & \cdot & \cdot & \\ & \cdot & \cdot & \cdot & \cdot & \cdot & \\ & \cdot & \cdot & \cdot & \cdot & \cdot & \\ \cdots & 0 & 2 & 0 & 0 & 0 & \cdots \\ \cdots & 0 & 2 & 0 & 0 & 0 & \cdots \\ \cdots & 2^{-1} & 1 & \boxed{2} & 2^2 & 2^3 & \cdots \end{pmatrix}^0 \longleftrightarrow 2F(z,w)$$

$$T(f;3,2) = \begin{pmatrix} \cdots & 0 & 0 & 0 & 1 & 0 & \cdots \\ \cdots & 0 & 0 & 0 & 1 & 0 & \cdots \\ \cdots & 2^{-4} & 2^{-3} & 2^{-2} & 2^{-1} & 1 & \cdots \\ \cdots & 0 & 0 & 0 & 0 & 0 & \cdots \\ \cdots & 0 & \boxed{0} & 0 & 0 & 0 & \cdots \end{pmatrix}^0 \leftrightarrow z^{-3}w^{-2}F(z,w)$$

$$N^2(f) = \begin{pmatrix} \cdots & 2^2 & 2 & 1 & 2^{-1} & 2^{-2} & \cdots \\ \cdots & 0 & 0 & 0 & 1 & 0 & \cdots \\ \cdots & 0 & 0 & 0 & 1 & 0 & \cdots \\ & \cdot & \cdot & \cdot & \cdot & \cdot & \\ & \cdot & \cdot & \cdot & \cdot & \cdot & \\ & \cdot & \cdot & \cdot & \cdot & \cdot & \end{pmatrix}^0 \leftrightarrow F(1/z,1/w)$$

Because of the isomorphism previously mentioned, there is no need to specify numerous equational constraints existing in $\Re[z,w]$. Indeed, for every identity involving ADD, $SCALAR$, $TRAN$, and $NINETY^2$ in $\Re^{Z\times Z}$ there is an identical identity involving $+$, $\cdot$, $z^{-n}w^{-m}$ and $\mathcal{I}$ respectively in $\Re[z,w]$. For instance the addition operation in $\Re[z,w]$ is associative, commutative, there is a zero infinite series and so on. Additionally, because range and domain induced operations always commute in $\Re^{Z\times Z}$ we have the "obvious" properties in $\Re[z,w]$.

P1) $z^{-n}w^{-m}(F(z,w)+G(z,w)) =$
$$z^{-n}w^{-m}F(z,w) + z^{-n}w^{-m}G(z,w)$$
P2) $z^{-n}w^{-m}(a(F(z,w))) = a(z^{-n}w^{-m}(F(z,w)))$
P3) $\mathcal{I}(F(z,w)+G(z,w)) = F(1/z,1/w)+G(1/z,1/w)$
P4) $\mathcal{I}(aF(z,w)) = aF(1/z,1/w)$

Example 4.7 *If*

$$f = \begin{pmatrix} 0 & 1 & 0 \\ 2 & \boxed{3} & 4 \\ 0 & 0 & 2 \end{pmatrix}^0 \quad \textit{and} \quad g = \begin{pmatrix} 3 & 2 & 1 \\ 0 & \boxed{5} & 0 \\ 0 & 1 & 0 \end{pmatrix}^0$$

then

$$F(z,w) = \mathcal{Z}(f) = 2z + 3 + 4z^{-1} + w^{-1} + 2z^{-1}w$$

and

$$G(z,w) = \mathcal{Z}(g) = 3zw^{-1} + 2w^{-1} + 5 + w + z^{-1}w^{-1}$$

Next observe that

$$N^2(f) + 3g_{1,0} = \begin{pmatrix} 2 & 9 & 6 & 3 \\ 4 & \boxed{3} & 17 & 0 \\ 0 & 1 & 3 & 0 \end{pmatrix}^0$$

This can be seen from the following block diagram walk through.

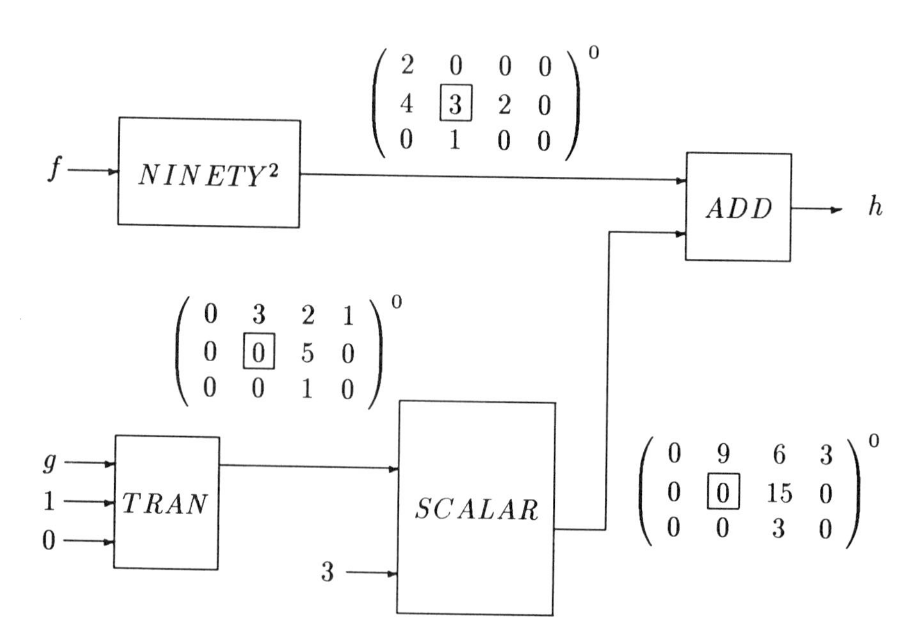

where

$$h = \begin{pmatrix} 2 & 9 & 6 & 3 \\ 4 & \boxed{3} & 17 & 0 \\ 0 & 1 & 3 & 0 \end{pmatrix}^{0}$$

The same result can be found using $\mathcal{Z}$ transforms. Indeed,

$$h = \mathcal{Z}^{-1}[F(1/z, 1/w) + 3z^{-1}G(z,w)]$$

Letting

$$F(1/z, 1/w) + 3z^{-1}G(z,w) = H(z,w)$$

gives

$$H(z,w) = 17z^{-1} + 3 + 4z + w + 2zw^{-1} + 9w^{-1}$$

$$+6z^{-1}w^{-1} + 3z^{-1}w + 3z^{-2}w^{-1}$$

Taking the inverse $\mathcal{Z}$ transform again gives h.

4.3 $\mathcal{Z}$ Transforms for Digital Signals of Finite Support

Each of the operations previously defined for $\mathcal{Z}$ transforms in general surely hold for $\mathcal{Z}$ transforms involving digital signals of finite support. As was seen in Example 4.7 of the last section, infinite sums is replaced by finite sums whenever digital signals are of finite support. Moreover, each of the four fundamental operations when applied to finite sums also result in this type of structure. It follows that the set of digital signals of finite support and the set of a finite series in $\Re[z,w]$ are isomorphic structures.

Example 4.8 *Suppose that*

$$f = \begin{pmatrix} 0 & -2 & 0 \\ 2 & -1 & 3 \end{pmatrix}^{0}_{-1,1}$$

and

$$g = \begin{pmatrix} 0 & 0 & 1 & 2 & 0 \\ 3 & 4 & 1 & 1 & 3 \end{pmatrix}^{0}_{-3,1}$$

then the corresponding $\mathcal{Z}$ transforms are

$$F(z,w) = 2z - 1 + 3z^{-1} - 2w^{-1}$$

and

$$G(z,w) = 3z^3 + 4z^2 + z + 1 + 3z^{-1} + zw^{-1} + 2w^{-1}$$

The $\mathcal{Z}$ transform of the sum of f and g, in the previous example, is a finite sum. The same is true for the $\mathcal{Z}$ transform of the scalar multiplication of f, the translation of f, and the 180° rotation of f. Specifically,

$$ADD(f,g) \longleftrightarrow 3z^3 + 4z^2 + 3z + 6z^{-1} + zw^{-1}$$

$$SCALAR(f;2) = 2f \longleftrightarrow 4z - 2 + 6z^{-1} - 4w^{-1}$$

$$TRAN(f;2,0) = f_{2,0} \longleftrightarrow 2z^{-1} - z^{-1} + 3z^{-3} - 2z^{-2}w^{-1}$$

and

$$NINETY^2(f) = N^2(f) \longleftrightarrow 3z - 1 + 2z^{-1} - 2w$$

One of the most important consequences of the isomorphism between the space of all digital signals of finite support and all finite $\mathcal{Z}$ transforms is the convolution theorem.

Theorem 4.1 *(Convolution Theorem) Suppose that f and g have finite support and*

$$f \longleftrightarrow F(z,w)$$

and

$$g \longleftrightarrow G(z,w)$$

then

$$f * g = \mathcal{Z}^{-1}(\mathcal{Z}(f)\mathcal{Z}(g))$$

Additionally, greater insight into convolution is achieved using this transform technique. The multiplication of the finite series in the Convolution Theorem should be carried out using the rules of "High school algebra" in conjunction with the operations given in the previous section. In other words, "multinomial multiplication" should be performed and like powers of z and w combined. Alternately, multiplication of these finite series can be found by the following simple, but forbidden looking procedure, if

$$F(z,w) = \sum_{j,k=0}^{\infty} a_{k,j} z^{-k-p} w^{-j-q}$$

and

$$G(z,w) = \sum_{j,k=0}^{\infty} b_{k,j} z^{-k-r} w^{-j-s}$$

where $a_{k,j} = 0$ for all $k > n$ and $j > m$. Similarly, $b_{k,j} = 0$ for all $n > n'$ and $j > m'$, then

$$F(z,w)G(z,w) = \sum_{j=0}^{m+m'-2} \sum_{k=0}^{n+n'-2} c_{k,j} z^{-(k+p+r)} w^{-(j+q+s)}$$

where

$$c_{k,j} = \sum_{t=0}^{j} \sum_{i=0}^{k} a_{i,t} b_{k-i,j-t}$$

Example 4.9 *Suppose that*

$$f = \begin{pmatrix} 3 & 0 \\ 2 & -1 \end{pmatrix}_{1,2}^{0}$$

and

$$g = \begin{pmatrix} -2 & 1 \\ 0 & 3 \end{pmatrix}_{3,4}^{0}$$

In Example 3.3 as well as subsequent examples, we saw that

$$f * g = \begin{pmatrix} -6 & 3 & 0 \\ -4 & 13 & -1 \\ 0 & 6 & -3 \end{pmatrix}_{4,6}^{0}$$

We will find $f * g$ utilizing the $\mathcal{Z}$ transform convolution theorem. To do this we first find the $\mathcal{Z}$ transform of f and g. Thus,

$$\mathcal{Z}(f) = F(z,w) = 3z^{-1}w^{-2} + 2z^{-1}w^{-1} - z^{-2}w^{-1}$$

and

$$\mathcal{Z}(g) = G(z,w) = -2z^{-3}w^{-4} + z^{-4}w^{-4} + 3z^{-4}w^{-3}$$

Next, we multiply these two finite sums, making sure that common terms are combined after the multiplication is concluded.

$$F(z,w)G(z,w) = (3z^{-1}w^{-2} + 2z^{-1}w^{-1} - z^{-2}w^{-1})(-2z^{-3}w^{-4} +$$

$$z^{-4}w^{-4} + 3z^{-4}w^{-3})$$

$$= -6z^{-4}w^{-6} - 4z^{-4}w^{-5} + 3z^{-5}w^{-6} + 13z^{-5}w^{-5}$$

$$-3z^{-6}w^{-4} + 6z^{-5}w^{-4} - z^{-6}w^{-5}$$

Finally,

$$f * g = \mathcal{Z}^{-1}(F(z,w)G(z,w))$$

and so

$$f * g = \begin{pmatrix} -6 & 3 & 0 \\ -4 & 13 & -1 \\ 0 & 6 & -3 \end{pmatrix}_{4,6}^{0}$$

as was previously noted.

A few observations with respect to the previous example are in order. One should compare the "High school" method of computing the product with the parallel algorithm for computing the convolution given in Figure 3.3. Indeed, the isomorphism becomes more apparent when the comparison is made. Additionally, the formula used at the end of Example 4.9 illustrates that the same results can be obtained in a tedious non-intuitive fashion.

To show that the convolution theorem holds, one need only utilize the isomorphism existing between finite duration signals and finite series in $\Re[z,w]$. Accordingly, we have

$$\mathcal{Z}(f * g) = \mathcal{Z}(ADD_{(n,m)\in \mathrm{coz}(g)}[f_{n,m} \cdot g(n,m)])$$

$$= \sum_{(n,m)\in \mathrm{coz}(g)} \mathcal{Z}(f_{n,m} \cdot g(n,m))$$

$$= \sum_{(n,m)\in \mathrm{coz}(g)} g(n,m)\, z^{-n}w^{-m}\, \mathcal{Z}(f)$$

$$= \mathcal{Z}(f) \sum_{(n,m)\in \mathrm{coz}(g)} g(n,m)\, z^{-n}w^{-m} = \mathcal{Z}(f) \cdot \mathcal{Z}(g)$$

The first equality above utilizes the parallel representation for convolution. The next equality makes use of property **Z1)** in section 4.2. The next equality involves property **Z2)** and **Z3)** also in section 4.2. The next equality holds since $\mathcal{Z}(f)$ dose not depend on the index of summation, and so it can be taken out of summation. The last equality holds using the definition of $\mathcal{Z}(g)$.

4.4 Laurent Expansions

As observed in the previous sections, a lot can be done with formal $\mathcal{Z}$ transforms and associated algebraic manipulations. However, it is often desirable to replace the indeterminates z and w by real, or more frequently by complex numbers. When this is done we must be cognizant on whether or not the series converges. The resulting series, in this case, we shall again denote by

$$F(z,w) = \sum_{n,m=-\infty}^{\infty} a_{n,m} z^{-n} w^{-m}$$

This series is called a Laurent expansion or Laurent series.

In one dimension, the Laurent series converges for complex values of z located in an annulus. That is, a ring shaped region of the form

$$r < \mid z \mid < R$$

or a degenerate annulus, that is a disk $\mid z \mid < R$ or outside of a disk $\mid z \mid > r$. Additionally, it may converge everywhere in the complex plane or nowhere. Thus, if any complex number in the region of convergence is substituted into the series, and the series is summed, a complex value is attained. Moreover, the convergence of the series is absolute and uniform on every compact subset within the annulus. Accordingly, the function $F(z)$ is analytic in the annulus type region.

In the two dimensional case, the series, when represented by an analytic function, will converge in a generalization of an annulus, called a Reinhart domain. Indeed, for $F(z,w)$ analytic in

$$r_1 < \mid z \mid < R_1 \quad \text{and} \quad r_2 < \mid w \mid < R_2$$

where

$$0 \leq r_1 < R_1 \leq \infty \quad \text{and} \quad 0 \leq r_2 < R_2 \leq \infty$$

$F(z,w)$ can be represented by the absolutely convergent series

$$\sum_{n,m=-\infty}^{\infty} a_{n,m} z^{-n} w^{-m}$$

The coefficients of the series are uniquely determined by F. Moreover, the series converges uniformly in any compact subset within the region given above. Recall that, because of the absolute convergence of the Laurent expansion within the Reinhart domain, the series can be summed, (that is arranged) in arbitrary order. The domain of convergence mentioned above is often illustrated as a rectangular region, see Figure 4.1.

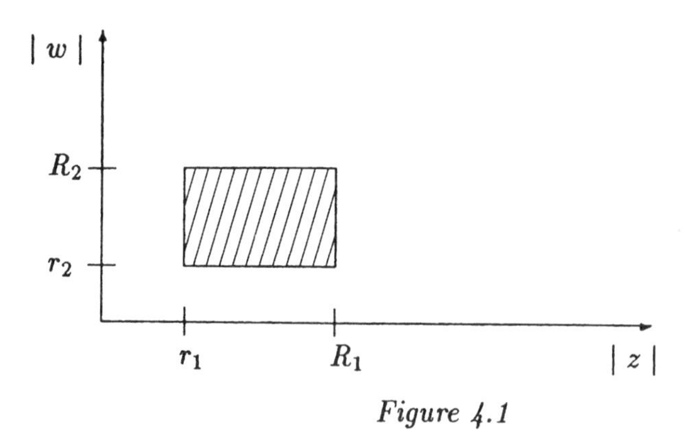

Figure 4.1

Notice that the one dimensional case reduces to a degenerate rectangle, namely the straight line

$$r_1 < \mid z \mid < R_1$$

Indeed, when we plot this inequality as a function of the real and imaginary parts of z it is an annulus. However, in terms of $\mid z \mid$ it is the degenerate rectangle mentioned above. Observe that if the analytic function $F(z,w)$ has a Laurent expansion convergent for

$$r_1 < \mid z \mid < R_1$$

and all w, then we obtain a region of convergence consisting of the semi-infinite strip region given in Figure 4.2.

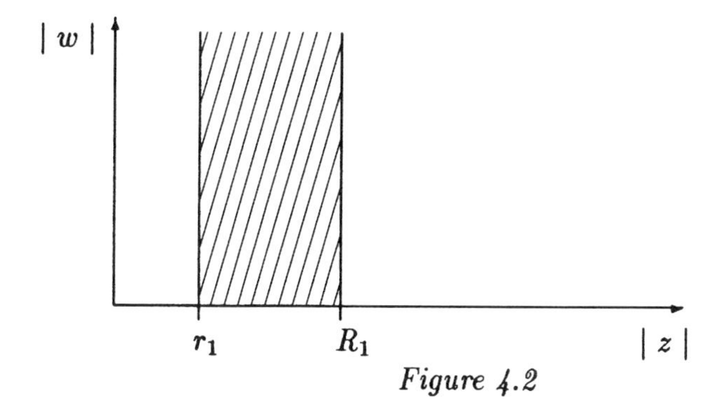

Figure 4.2

For nonzero digital signals of finite support, their Laurent expansion will converge absolutely everywhere, except possibly for $z = 0$ or $w = 0$. Several possibilities arise:

1. The Laurent expansion may be valid everywhere. This is true for signals with support in the third quadrant.

2. The Laurent expansion may be valid everywhere except possibly $z = 0$. This is true for signals f, such that $\mathrm{coz}(f)$ contains points only from the second and third quadrant.

3. The Laurent expansion may be valid everywhere except possibly $w = 0$. This is true for signals f such that $\mathrm{coz}(f)$ contains points only from the third and fourth quadrants.

4. The Laurent expansion may be valid everywhere except possibly $z = 0$ and $w = 0$. This is true for signals f such that $\mathrm{coz}(f)$ contains points from the second, third and fourth quadrant.

Example 4.10 *Several digital signals of finite support are given along with their Laurent expansions.*

1.

$$f = \begin{pmatrix} 1 & 2 \\ 0 & 3 \end{pmatrix}_{2,3}^{0} \leftrightarrow z^{-2}w^{-3} + 2z^{-3}w^{-3} + 3z^{-3}w^{-2} = F(z,w)$$

$F(z,w)$ is valid everywhere except $z = 0$ or $w = 0$.

2.

$$g = \begin{pmatrix} -2 & 1 \\ 3 & 0 \end{pmatrix}_{-3,3}^{0} \leftrightarrow -2z^3w^{-3} + z^2w^{-3} + 3z^3w^{-3} = G(z,w)$$

$G(z,w)$ is valid everywhere except $w = 0$.

3.

$$h = \begin{pmatrix} 2 & -4 \\ 1 & 0 \end{pmatrix}_{-5,-6}^{0} \leftrightarrow 2z^5w^6 + 2z^5w^7 - 4z^4w^6 = H(z,w)$$

$H(z,w)$ is valid everywhere.

4.

$$k = \begin{pmatrix} 3 & -1 \\ 1/2 & 0 \end{pmatrix}_{4,-3}^{0} \leftrightarrow 3z^{-4}w^3 + \frac{1}{2}z^{-4}w^4 - z^{-5}w^3 = K(z,w)$$

$K(z,w)$ is valid everywhere except at $z = 0$.

The Reinhart domain associated with signals with support in a single quadrant can be described in a simpler manner than arbitrary digital signals. Perhaps, the simplest Reinhart domain results from signals f with support in the third quadrant. Here we can write:

$$f = \begin{pmatrix} \cdots & f(-1,0) & \boxed{f(0,0)} \\ \cdots & f(-1,-1) & f(0,-1) \\ & \cdot & \cdot \\ & \cdot & \cdot \\ & \cdot & \cdot \end{pmatrix}^{0}$$

Accordingly,

$$\mathcal{Z}(f) = F(z,w) = \sum_{j,k=-\infty}^{0} f(k,j)\, z^{-k}w^{-j} = \sum_{j,k=0}^{\infty} f(-k,-j)\, z^k w^j$$

Hence, these types of signals are represented by two dimensional power series. This is the simplest type of possibly non-finite Laurent expansion.

It is known that if some arrangement of the series converges for some values

$$z = z_0 \neq 0 \quad \text{and} \quad w = w_0 \neq 0$$

then this series converges absolutely for all z and w such that

$$|\, z \,| < z_0, \quad |\, w \,| < w_0$$

Moreover, the function $F(z, w)$ will be analytic in this region. This region is illustrated in Figure 4.3. The $\mathcal{Z}$ transform, in this case, is the Taylor series and the series for $F(z, w)$ converges uniformly to $F(z, w)$ in any compact set within this Reinhart domain.

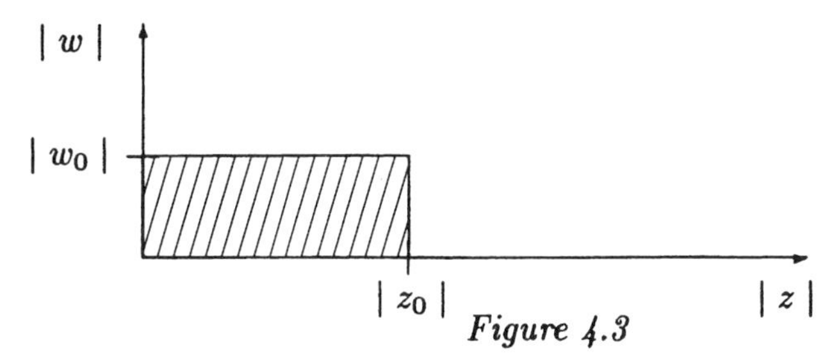

Figure 4.3

A consequence of the above result is that the region of absolute convergence for the $\mathcal{Z}$ transform of a third quadrant signal must always contain the interior of rectangles.

Example 4.11 *Consider the third quadrant digital signal*

$$f = \begin{pmatrix} \cdots & 0 & 0 & 0 & 0 & 0 & 0 & \boxed{0} \\ \cdots & 0 & 0 & 0 & 0 & 1 & 0 & 0 \\ \cdots & 0 & 0 & 1 & 0 & 0 & 0 & 0 \\ & 1 & 0 & 0 & 0 & 0 & 0 & 0 \\ & \cdot & \cdot & \cdot & \cdot & \cdot & \cdot & \cdot \\ & \cdot & \cdot & \cdot & \cdot & \cdot & \cdot & \cdot \\ & \cdot & \cdot & \cdot & \cdot & \cdot & \cdot & \cdot \end{pmatrix}^{0} \leftrightarrow \sum_{n=0}^{\infty} z^{2n} w^{n} = F(z, w)$$

Then the power series converges absolutely for $z = 0$ and all w; it converges absolutely for $w = 0$ and all z, and it also converges absolutely for

$$|\, z^2 w \,| < 1 \quad \text{or} \quad |\, w \,| < \frac{1}{|\, z \,|^2}$$

where $z \neq 0$ and $w \neq 0$. This region is plotted in Figure 4.4. *Observe that for any point within this region, the rectangular box with sides parallel to the coordinate axes is also in the region.*

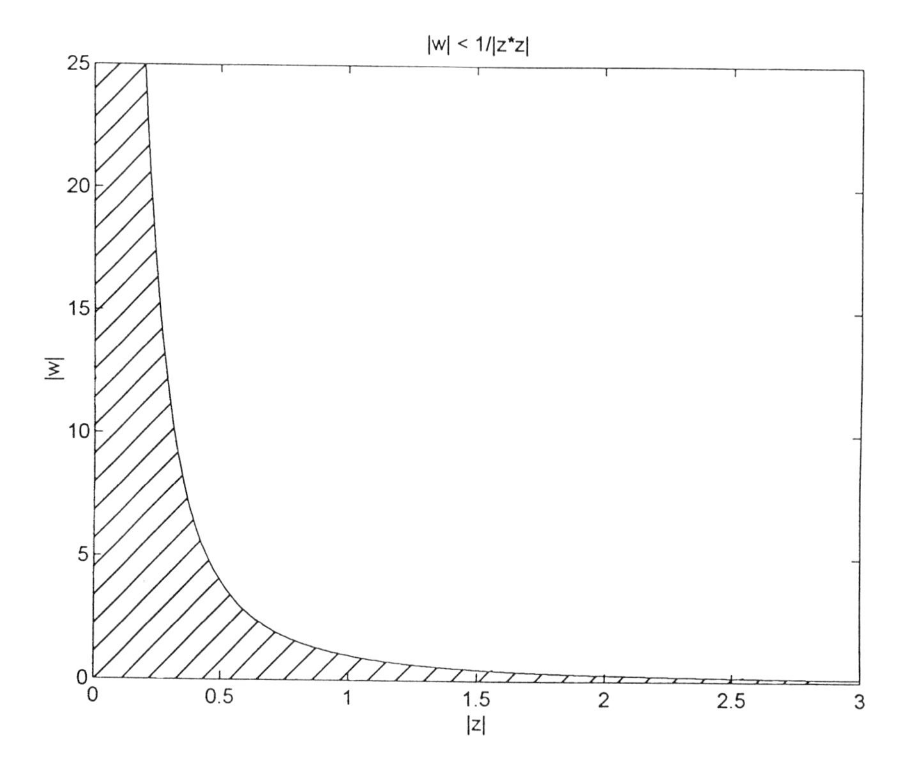

Figure 4.4

Example 4.12 *Suppose that*

$$f = \begin{pmatrix} \cdots & 1 & 1 & 1 & \boxed{1} \\ \cdots & & 6 & 4 & 2 \\ \cdots & & & 12 & 4 \\ & & & \cdot & 8 \\ & & & \cdot & \cdot \\ & & & \cdot & \cdot \end{pmatrix}^{0} \leftrightarrow \sum_{j,k=0}^{\infty} f(-k,-j) z^k w^j = F(z,w)$$

The test for absolute convergence entails

$$\sum_{j,k=0}^{\infty} |f(-k,-j)| \, |z|^k \, |w|^j$$

When utilizing points within the region of absolute convergence, the ordering of terms will not affect the outcome. Thus, we can write the absolutely convergent series as

$$\sum_{n=0}^{\infty} (|z| + 2|w|)^n = 1 + (|z| + 2|w|) + (|z|^2 + 4|z|\,|w| + 4|w|^2)$$

$$+ (|z|^3 + 6|z|^2|w| + 12|z|\,|w|^2 + 8|w|^3) + \ldots$$

Accordingly, the series converges when

$$|z| + 2|w| < 1$$

This triangular region is plotted in Figure 4.5.

The Reinhart region, illustrated in Figure 4.5, can also be described using logarithms. In this case, for each point

$$(|z|, |w|), \; z \neq 0, \; w \neq 0$$

in the Reinhart domain we let

$$t = \ln|z| \quad \text{and} \quad s = \ln|w|$$

What is of consequence is that the logarithmic image determined above will always be convex. That is, for any two points in this image, the straight line segment connecting these two points will always be in the set. The corresponding logarithm domain for Example 4.12 above is depicted in Figure 4.6.

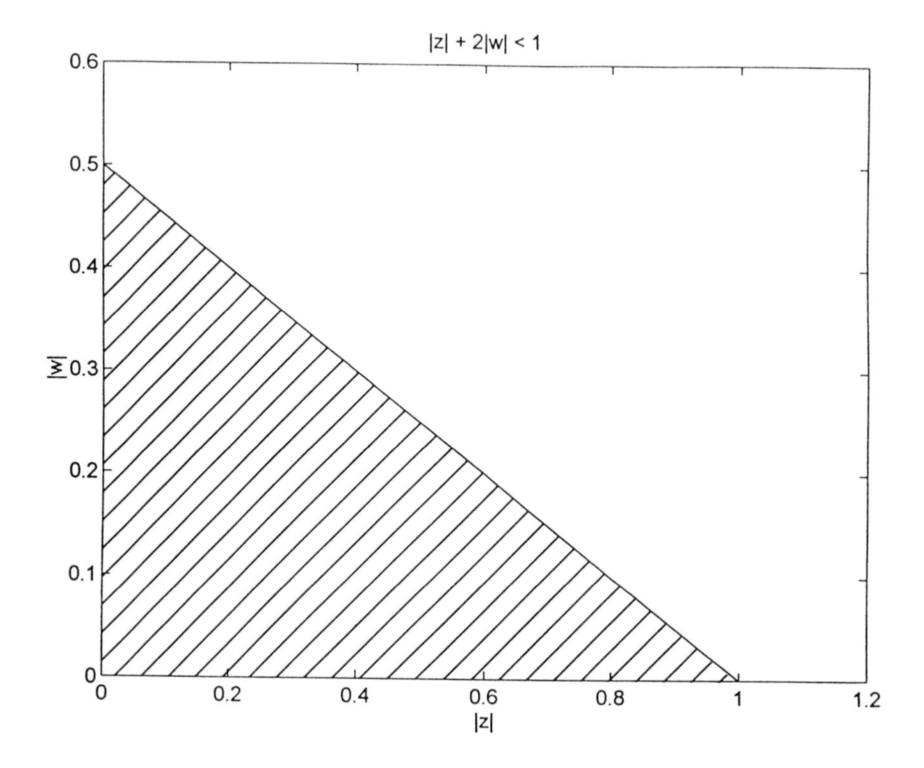

Figure 4.5

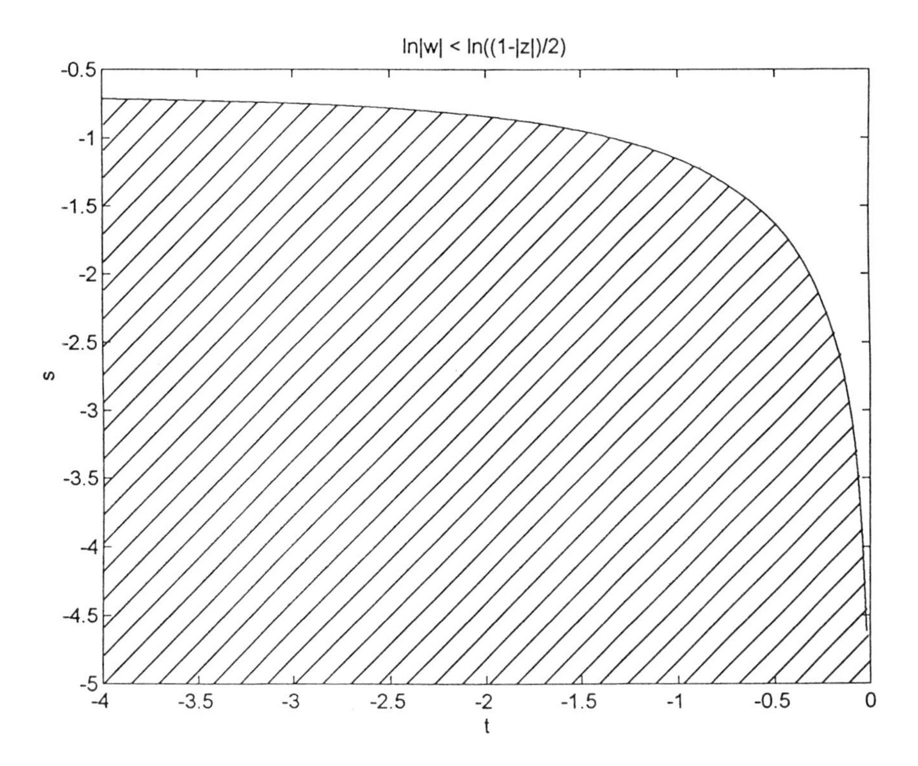

Figure 4.6

For signals f with support in the first quadrant

$$f = \begin{pmatrix} \vdots & \vdots & \\ f(0,1) & f(1,1) & \cdots \\ \boxed{f(0,0)} & f(1,0) & \cdots \end{pmatrix}^0$$

The $\mathcal{Z}$ transform $F(z,w)$ consists of the constant $f(0,0)$ term plus a principal part of the Laurent expansion. Indeed,

$$F(z,w) = \sum_{m,n=0}^{\infty} f(n,m) z^{-n} w^{-m}$$

Similar to the power series situation, if this series converges at (z_0, w_0), it will converge absolutely for all points (z,w) such that

$$|z| \geq |z_0| \quad \text{and} \quad |w| \geq |w_0|$$

The resulting function will be analytic in this region, and again logarithms can be utilized in describing the region of absolute convergence. Here we again let $t = \ln|z|$ and $s = \ln|w|$ and the logarithmic region thereby obtained will again be convex.

Example 4.13 *Let*

$$f = \begin{pmatrix} \cdot & \cdot & \cdot & \cdot & \cdot & \\ \cdot & \cdot & \cdot & \cdot & \cdot & \\ \cdot & \cdot & \cdot & \cdot & \cdot & \\ 0 & 0 & 0 & 0 & 16 & \cdots \\ 0 & 0 & 0 & 0 & 0 & \cdots \\ 0 & 0 & 4 & 0 & 0 & \cdots \\ 0 & 0 & 0 & 0 & 0 & \cdots \\ \boxed{1} & 0 & 0 & 0 & 0 & \cdots \end{pmatrix}^0 \leftrightarrow \sum_{n=0}^{\infty} \left(2z^{-1}w^{-1}\right)^{2n} = F(z,w)$$

This series converges absolutely for

$$|z| > \frac{2}{|w|}$$

The Reinhart domain is illustrated in Figure 4.7 *and the corresponding convex logarithmic domain is illustrated in Figure* 4.8.

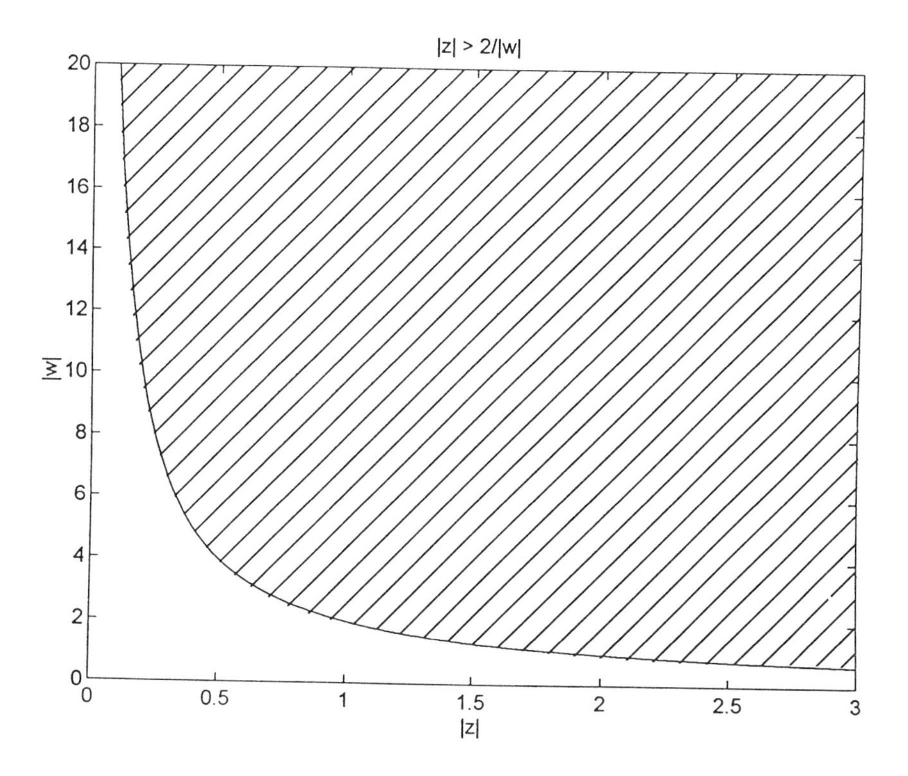

Figure 4.7

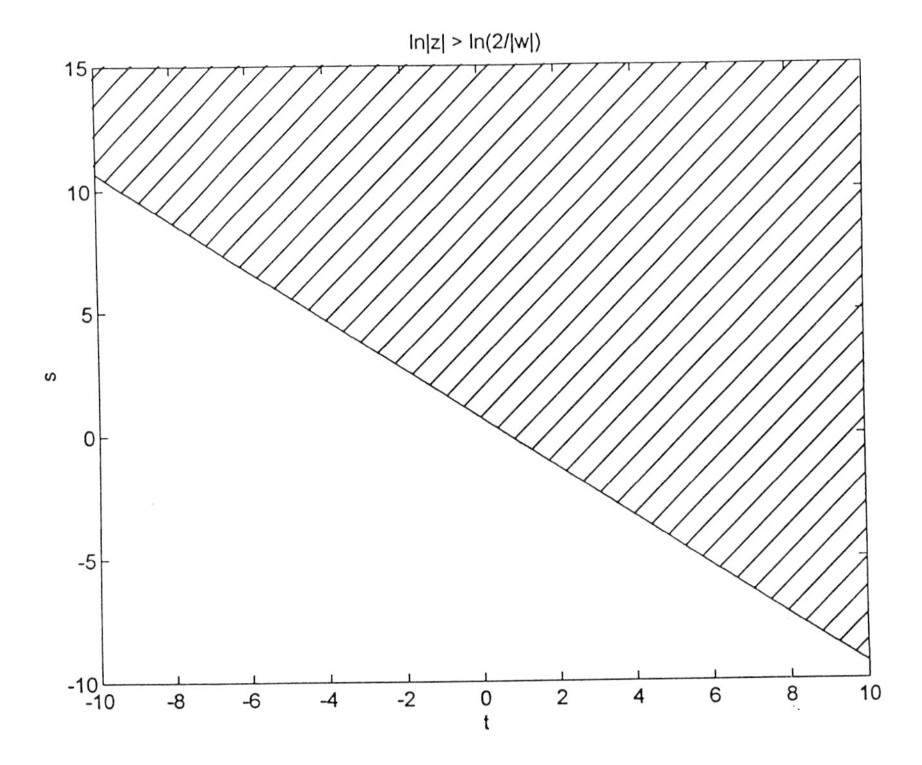

Figure 4.8

Similar descriptions can be given for regions of absolute convergence for signals with support in the second and fourth quadrant.

The Laurent expansion arising from the $\mathcal{Z}$ transform of an arbitrary digital signal can converge nowhere, everywhere or somewhere. In the latter case, the region of convergence can be found by writing the original signal as a sum of four single quadrant signals. Then, find the Reinhart domain associated with each simple quadrant signal. Finally, by forming the intersection of these domains we obtain the region for which the original signal converges.

Example 4.14 *Suppose that we are given the "diagonal valued digital signal f" where*

$$f(n,m) = \begin{cases} 1/2^n & n = m > 0 \\ 1 & n = m < 0 \\ 0 & otherwise \end{cases}$$

then

$$f \longleftrightarrow F(z,w) = \sum_{n,m=1}^{\infty} z^n w^m + \sum_{n,m=1}^{\infty} 2^{-n} z^{-n} w^{-m}$$

the first series above converges absolutely for $z \neq 0$, $w \neq 0$ and

$$|zw| < 1 \text{ or } |w| < 1/|z|$$

The second series converges absolutely for

$$1/(2|z|) < |w|$$

Thus, $F(z,w)$ converges in the region

$$1/(2|z|) < |w| < 1/|z|$$

Before we conclude this section, it should mentioned that an arbitrary analytic function $F(z,w)$ might have many or even an infinite number of Laurent expansions associated with it. Each will be valid in some Reinhart domain. Moreover, there will be a unique digital signal for which the corresponding $\mathcal{Z}$ transform will be the Laurent expansion valid in that Reinhart domain.

Example 4.15 *If*

$$F(z,w) = 1/(1 - zw)$$

then for

$$|zw| < 1 \text{ or } |w| < 1/|z|$$

we have

$$F(z,w) = 1 + zw + z^2w^2 + \cdots$$

and here

$$f \longleftrightarrow F(z,w)$$

Additionally,

$$f = \begin{pmatrix} \cdots & 0 & 0 & \boxed{1} \\ \cdots & 0 & 1 & 0 \\ \cdots & 1 & 0 & 0 \\ & \cdot & \cdot & \cdot \\ & \cdot & \cdot & \cdot \\ & \cdot & \cdot & \cdot \end{pmatrix}^0$$

On the other hand,

$$F(z,w) = \frac{1}{(1 - zw)} = \frac{-1}{zw} \cdot \frac{1}{1 - \frac{1}{zw}}$$

$$= \frac{-1}{zw}(1 + \frac{1}{zw} + \frac{1}{z^2w^2} + \cdots)$$

This is valid where

$$1/|zw| < 1 \text{ or } |w| > 1/|z|$$

and so

$$f \longleftrightarrow F(z,w)$$

where

$$f = \begin{pmatrix} \cdot & \cdot & \cdot & \\ \cdot & \cdot & \cdot & \\ \cdot & \cdot & \cdot & \\ 0 & 0 & 1 & \cdots \\ 0 & 1 & 0 & \cdots \\ \boxed{0} & 0 & 0 & \cdots \end{pmatrix}^0$$

4.5 $\mathcal{Z}$ Transform for l_1 Signals and Signals of Arbitrary Support

In section 3.5 we saw that any two signals in l_1 can be convolved and the result is also in l_1. Not all l_1 signals, however, have $\mathcal{Z}$ transforms representing analytic functions.

Example 4.16 *Suppose that the l_1 signal*

$$f = (\cdots \frac{1}{3^2}\ \frac{1}{2^2}\ \boxed{1}\ \frac{1}{2^2}\ \frac{1}{3^2}\ \cdots)^0$$

then formally, the $\mathcal{Z}$ transform of f is $F(z,w)$ where

$$F(z,w) = \sum_{n=1}^{\infty} \frac{1}{n^2} z^{-n} + 1 + \sum_{n=-\infty}^{-1} \frac{1}{n^2} z^n$$

This series converges for $|z| = 1$ and all w. However, this expansion does not converge in any domain. Accordingly, we can only manipulate the series in a formal matter, or use it for complex values z such that $|z| = 1$ and arbitrary w.

The results presented in the previous example typify l_1 signals — they always have $\mathcal{Z}$ transforms which converges (absolutely) at least for $|z| = |w| = 1$. For signals with support in the first quadrant

$$f = \begin{pmatrix} & \cdot & \cdot & \cdot & \cdot & \cdot & \\ & \cdot & \cdot & \cdot & \cdot & \cdot & \\ & \cdot & \cdot & \cdot & \cdot & \cdot & \\ \cdots & 0 & 0 & a_{0,1} & a_{1,1} & a_{2,1} & \cdots \\ \cdots & 0 & 0 & \boxed{a_{0,0}} & a_{1,0} & a_{2,0} & \cdots \\ \cdots & 0 & 0 & 0 & 0 & 0 & \cdots \\ & \cdot & \cdot & \cdot & \cdot & \cdot & \\ & \cdot & \cdot & \cdot & \cdot & \cdot & \\ & \cdot & \cdot & \cdot & \cdot & \cdot & \end{pmatrix}^0$$

have $\mathcal{Z}$ transforms which represent analytic functions for all $|z|, |w| \geq 1$. It may converges in even larger regions and therefore represent an analytic function in the larger region.

Example 4.17 *Consider the first quadrant signal*

$$f = \begin{pmatrix} \cdot & \cdot & \cdot & \cdot & \\ \cdot & \cdot & \cdot & \cdot & \\ \cdot & \cdot & \cdot & \cdot & \\ 0 & 0 & 0 & 1/3^2 & \cdots \\ 0 & 0 & 1/2^2 & 0 & \cdots \\ 0 & 1 & 0 & 0 & \cdots \\ \boxed{0} & 0 & 0 & 0 & \cdots \end{pmatrix}^0$$

The $\mathcal{Z}$ transform of f is

$$F(z, w) = \sum_{n=1}^{\infty} \frac{1}{n^2} z^{-n} w^{-n}$$

The series converges (absolutely) for all $|zw| \geq 1$.

The convolution theorem also holds for signals in l_1.

Theorem 4.2 *(Convolution Theorem) Let f and g be in l_1 with*

$$f \longleftrightarrow F(z, w)$$
$$g \longleftrightarrow G(z, w)$$

then

$$f * g \longleftrightarrow F(z, w)G(z, w)$$

As a consequence, convolution can be found by taking $\mathcal{Z}$ transforms of f and g, multiplying these quantities and taking the inverse $\mathcal{Z}$ transform of the result, that is

$$f * g = \mathcal{Z}^{-1}[\mathcal{Z}(f)\mathcal{Z}(g)]$$

The multiplication of two $\mathcal{Z}$ transforms of functions in l_1 is well defined. It converges absolutely, at least in the intersection of the regions of absolute convergence of $\mathcal{Z}(f)$ and $\mathcal{Z}(g)$, which is at least $|z| = |w| = 1$. If

$$\mathcal{Z}(f) = \sum_{m,n=-\infty}^{\infty} a_{n,m} z^{-n} w^{-m}$$

$$\mathcal{Z}(g) = \sum_{m,n=-\infty}^{\infty} b_{n,m} z^{-n} w^{-m}$$

then the multiplication results in

$$\mathcal{Z}(f)\mathcal{Z}(g) = \sum_{m,n=-\infty}^{\infty} c_{n,m} z^{-n} w^{-m}$$

where

$$c_{n,m} = \sum_{i,j=-\infty}^{\infty} a_{i,j} b_{n-i,m-j}$$

Using the isomorphism in section 4.2 gives the association

$$c_{n,m} = (f * g)(n,m) = \sum_{i,j=-\infty}^{\infty} f(i,j) g(n-i, m-j)$$

which proves the convolution theorem in this case. We will conclude this section with another important case in which the convolution theorem holds.

Theorem 4.3 *(Convolution Theorem) If f is in $\Re^{Z \times Z}$ and g is of finite support then we can formally find the $\mathcal{Z}$ transform $F(z,w)$ of f. If $G(z,w)$ is the $\mathcal{Z}$ transform of g, then*

$$f * g \longleftrightarrow F(z,w)G(z,w)$$

Example 4.18 *Let us illustrate the convolution theorem using*

$$f = 1_{Z \times Z}$$

and

$$g = \begin{pmatrix} 1 & 0 \\ 2 & 3 \end{pmatrix}_{0,1}^{0}$$

Since

$$f \longleftrightarrow F(z,w) = \sum_{m,n=-\infty}^{\infty} z^{-n} w^{-m}$$

and

$$g \longleftrightarrow G(z,w) = 2 + 3z^{-1} + w^{-1}$$

we have

$$F(z,w)G(z,w) = (\sum_{m,n=-\infty}^{\infty} z^{-n}w^{-m})(2 + 3z^{-1} + w^{-1})$$

$$= \sum_{m,n=-\infty}^{\infty} 2z^{-n}w^{-m} + \sum_{m,n=-\infty}^{\infty} 3z^{-n-1}w^{-m} + \sum_{m,n=-\infty}^{\infty} z^{-n}w^{-m-1}$$

$$= \sum_{m,n=-\infty}^{\infty} 6z^{-n}w^{-m}$$

therefore,

$$f * g = \mathcal{Z}^{-1}(\sum_{m,n=-\infty}^{\infty} 6z^{-n}w^{-m}) = 6_{Z\times Z}$$

4.6 The $\mathcal{Z}$ Transform for Calculating Correlation

Since convolution and correlation differ only by a 180° rotation, we should expect that the $\mathcal{Z}$ transform can be used to obtain the correlation of signals. Indeed, this is the case for various classes of signals in $\Re^{Z\times Z}$. Again, the isomorphism presented earlier provides the key to proving the result.

Theorem 4.4 *(Correlation Theorem) Suppose that f and g are in l_1, or that f is any digital signal and g is of finite support. If*

$$f \longleftrightarrow F(z,w)$$

$$g \longleftrightarrow G(z,w)$$

then

$$COR(f,g) = \mathcal{Z}^{-1}(F(z,w)G(1/z,1/w))$$

The procedure is illustrated in Figure 4.9.

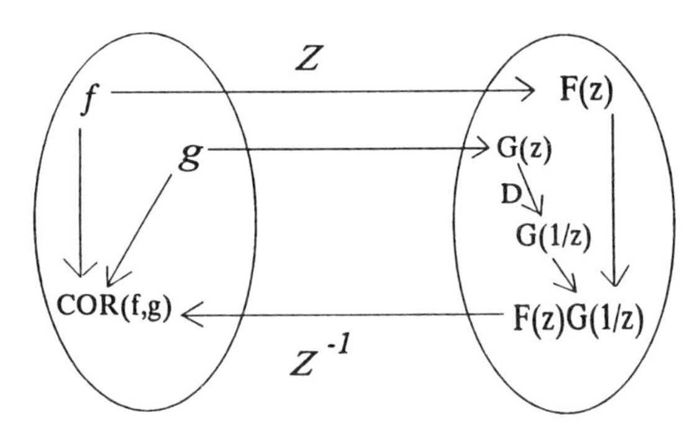

Figure 4.9

Thus, to find the correlation of f and g, the $\mathcal{Z}$ transform of each signal is found. The inversion operation is applied to the $\mathcal{Z}$ transform of g. The product is formed and an inverse transform is taken.

Example 4.19 *In Example* 3.16, *we used the signals*

$$f = \begin{pmatrix} 2 & 1 \\ -1 & 3 \end{pmatrix}_{0,0}^{0}$$

$$g = \begin{pmatrix} 0 & 1 \\ -1 & 3 \end{pmatrix}_{0,0}^{0}$$

and found

$$COR(f,g) = \begin{pmatrix} 6 & 1 & -1 \\ -1 & 11 & -3 \\ -1 & 3 & 0 \end{pmatrix}_{-1,1}^{0}$$

Let us again find this result, this time using correlation theorem. We have

$$f \longleftrightarrow F(z,w) = 2 + z^{-1} - w + 3z^{-1}w$$

and

$$g \longleftrightarrow G(z,w) = z^{-1} - w + 3z^{-1}w$$

First forming

$$G(1/z, 1/w) = z - w^{-1} + 3zw^{-1}$$

then multiplying $F(z,w)$ *with* $G(1/z,1/w)$ *gives*

$$F(z,w)G(1/z,1/w) = 11+w^{-1}+6zw^{-1}-z^{-1}w^{-1}-z-3z^{-1}-zw+3w$$

Taking the inverse $\mathcal{Z}$ transform gives

$$COR(f,g) = \begin{pmatrix} 6 & 1 & -1 \\ -1 & 11 & -3 \\ -1 & 3 & 0 \end{pmatrix}_{-1,1}^{0}$$

4.7 Transfer Functions

In section 3.7 it was seen how convolution is used in shaping signals by filtering. In this application even though convolution involves two arguments, f and h, one of these is thought to be the input signal. Here we use f as the input signal. The other signal, h is the structuring signal. It is found using heuristics. When h is applied to f (via convolution), the response (output signal) g usually preserves or magnifies "good" characteristics of f and attenuates "bad" characteristics of f.

If we let $F(z,w)$ be the $\mathcal{Z}$ transform of the input signal and $G(z,w)$ be the $\mathcal{Z}$ transform of the output signal, then the transfer function is denoted by $H(z,w)$ and is defined to be

$$H(z,w) = \frac{G(z,w)}{F(z,w)}$$

In this discussion, $F(z,w)$ is often the excitation transform, $G(z,w)$ is called the response transform, and $H(z,w)$ is also called the system function. The process described above is depicted by he block diagram

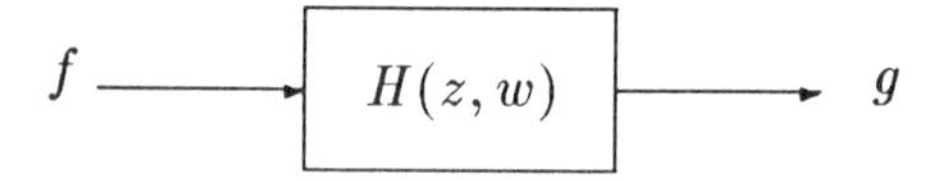

Example 4.20 *Among the most simple, yet important examples of a transfer function is the one which produces a delay or shift to the right by one unit. It has input*

$$f = \begin{pmatrix} & \vdots & \vdots & \vdots & \vdots & \\ \cdots & a_{-1,1} & a_{0,1} & a_{1,1} & a_{2,1} & \cdots \\ \cdots & a_{-1,0} & \boxed{a_{0,0}} & a_{1,0} & a_{2,0} & \cdots \\ \cdots & a_{-1,-1} & a_{0,-1} & a_{1,-1} & a_{2,-1} & \cdots \\ & \vdots & \vdots & \vdots & \vdots & \end{pmatrix}$$

and has output

$$g = \begin{pmatrix} & \vdots & \vdots & \vdots & \vdots & \\ \cdots & a_{-2,1} & a_{-1,1} & a_{0,1} & a_{1,1} & \cdots \\ \cdots & a_{-2,0} & \boxed{a_{-1,0}} & a_{0,0} & a_{1,0} & \cdots \\ \cdots & a_{-2,-1} & a_{-1,-1} & a_{0,-1} & a_{1,-1} & \cdots \\ & \vdots & \vdots & \vdots & \vdots & \end{pmatrix}$$

Since

$$f \longleftrightarrow F(z,w) = \sum_{m,n=-\infty}^{\infty} a_{n,m} z^{-n} w^{-m}$$

and

$$g \longleftrightarrow G(z,w) = \sum_{m,n=-\infty}^{\infty} a_{n,m} z^{-n-1} w^{-m} = \frac{1}{z} \sum_{m,n=-\infty}^{\infty} a_{n,m} z^{-n} w^{-m}$$

the transfer function is

$$H(z,w) = \frac{G(z,w)}{F(z,w)} = \frac{1}{z}$$

This is illustrated by the familiar block diagram

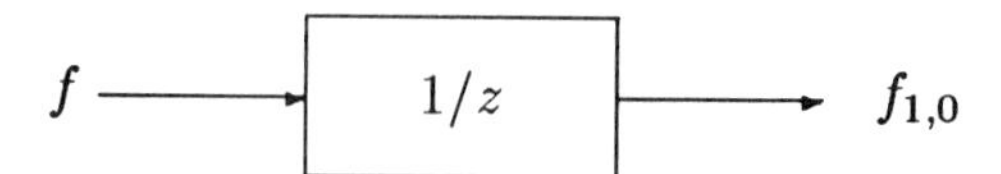

A similar argument can be made for a shift in the upward direction. Indeed, the resulting transfer function in this case is $1/w$ and the block diagram is given by

$$f \longrightarrow \boxed{1/w} \longrightarrow f_{0,1}$$

Example 4.21 *Let us assume that the input signal is*

$$f = \begin{pmatrix} 1 & 2 & 3 \\ 2 & -1 & 0 \\ 1 & 2 & 3 \end{pmatrix}_{0,2}^{0}$$

and the output signal is

$$g = \begin{pmatrix} 0 & 1/5 & 2/5 & 3/5 & 0 \\ 1/5 & 1 & 1 & 1 & 3/5 \\ 2/5 & 3/5 & 1 & 1 & 0 \\ 1/5 & 1 & 1 & 1 & 3/5 \\ 0 & 1/5 & 2/5 & 3/5 & 0 \end{pmatrix}_{-1,3}^{0}$$

The objective is to find the transfer function $H(z,w)$. Since

$$F(z,w) = w^{-2}+2z^{-1}w^{-2}+3z^{-2}w^{-2}+2w^{-1}-z^{-1}w^{-1}+1+2z^{-1}+3z^{-2}$$

$$G(z,w) = \frac{1}{5}w^{-3} + \frac{2}{5}z^{-1}w^{-3} + \frac{3}{5}z^{-2}w^{-3}$$
$$+\frac{1}{5}zw^{-2} + w^{-2} + z^{-1}w^{-2} + z^{-2}w^{-2} + \frac{3}{5}z^{-3}w^{-2}$$
$$+\frac{2}{5}zw^{-1} + \frac{3}{5}w^{-1} + z^{-1}w^{-1} + z^{-2}w^{-1}$$
$$+\frac{1}{5}z + 1 + z^{-1} + z^{-2} + \frac{3}{5}z^{-3}$$

$$+\frac{1}{5}w + \frac{2}{5}z^{-1}w + \frac{3}{5}z^{-2}w.$$

Consequently,

$$H(z,w) = \frac{G(z,w)}{F(z,w)} = \frac{1}{5}(w^{-1} + z + 1 + z^{-1} + w)$$

This is the $\mathcal{Z}$ transform of the 5 point moving average filter involving values at lattice point and its strong neighbors. Hence, the structuring signal or impulse response is

$$h = \begin{pmatrix} 0 & 1/5 & 0 \\ 1/5 & \boxed{1/5} & 1/5 \\ 0 & 1/5 & 0 \end{pmatrix}^{0}$$

This process is depicted using a block diagram as

$$f \longrightarrow \boxed{1/5(zw + z^{-1}w + 1 + zw^{-1} + z^{-1}w^{-1})} \longrightarrow g$$

The transfer function here was found by dividing $F(z,w)$ into $G(z,w)$ using long division. The result can be verified by multiplication. In general, the transfer function need not be expressible as a finite sum.

Example 4.22 *Suppose that the input signal is*

$$f = \begin{pmatrix} 1 & -1 \end{pmatrix}^{0}_{0,0}$$

and the output signal is

$$g = \begin{pmatrix} 0 & -1 \\ 1 & 0 \end{pmatrix}^{0}_{0,1}$$

we will find the transfer function $H(z,w)$ for this process. Here

$$F(z,w) = 1 - z^{-1}$$

$$G(z,w) = 1 - z^{-1}w^{-1}$$

Hence

$$H(z,w) = \frac{1 - z^{-1}w^{-1}}{1 - z^{-1}}$$

It is instructive to determine the Laurent series associated with $H(z,w)$. This enables us to determine the structuring signal (impulse response)

$$h = \mathcal{Z}^{-1}(H(z,w)).$$

Since for $|z| > 1$,

$$\frac{1}{1 - z^{-1}} = 1 + z^{-1} + z^{-2} + \cdots$$

we have

$$H(z,w) = (1 - z^{-1}w^{-1})(1 + z^{-1} + z^{-2} + \cdots)$$

$$= 1 - z^{-1}w^{-1} + z^{-1} - z^{-2}w^{-1} + z^{-2} - z^{-3}w^{-1} + \cdots$$

Hence, the structuring signal is

$$h = \mathcal{Z}^{-1}(H(z,w)) = \begin{pmatrix} \cdot & \cdot & \cdot & \cdot & \\ \cdot & \cdot & \cdot & \cdot & \\ \cdot & \cdot & \cdot & \cdot & \\ 0 & 0 & 0 & 0 & \cdots \\ 0 & -1 & -1 & -1 & \cdots \\ \boxed{1} & 1 & 1 & 1 & \cdots \end{pmatrix}^{0}$$

On the other hand, manipulating the transfer function into the form

$$H(z,w) = \frac{w^{-1} - z}{1 - z}$$

and noticing that

$$\frac{1}{z} = 1 + z + z^2 + \cdots$$

for $|z| < 1$.

$$H(z,w) = (w^{-1} - z)(1 + z + z^2 + \cdots)$$

$$= w^{-1} - z + zw^{-1} - z^2 + z^2w^{-1} - z^3 + \cdots$$

accordingly, another impulse response is given by

$$h = \mathcal{Z}^{-1}(H(z,w)) = \begin{pmatrix} \cdots & 1 & 1 & 1 & \boxed{1} \\ \cdots & -1 & -1 & -1 & 0 \\ \cdots & 0 & 0 & 0 & 0 \\ & \cdot & \cdot & \cdot & \cdot \\ & \cdot & \cdot & \cdot & \cdot \\ & \cdot & \cdot & \cdot & \cdot \end{pmatrix}^{0}$$

Thus, there need not be a unique solution.

One could verify the result above by actually convolving the input signal f with structuring signal h to obtain g.

4.8 Exercises:

1. Due to the one to one correspondence between $\Re^{Z\times Z}$ and the set of all $\mathcal{Z}$ transforms, every bound matrix f in $\Re^{Z\times Z}$ has a unique $\mathcal{Z}$ transform $F(z,w)$. Conversely, every $\mathcal{Z}$ transform $G(z,w)$ has a unique digital signal g in $\Re^{Z\times Z}$ for which $G(z,w)$ is its $\mathcal{Z}$ transform.

   (a) If

   $$f = \begin{pmatrix} 2 & 0 & 0 & 0 & 1 \\ 3 & 2 & 0 & 1 & 4 \end{pmatrix}^{0}_{-1,-1}$$

   find $F(z,w)$.

   (b) If

   $$G(z,w) = z^3 - z^{-2} + wz$$

   find g.

2. For the digital signals f and g given Exercise 4.1 determine the $\mathcal{Z}$ transform for the output of the following blocks

   a)

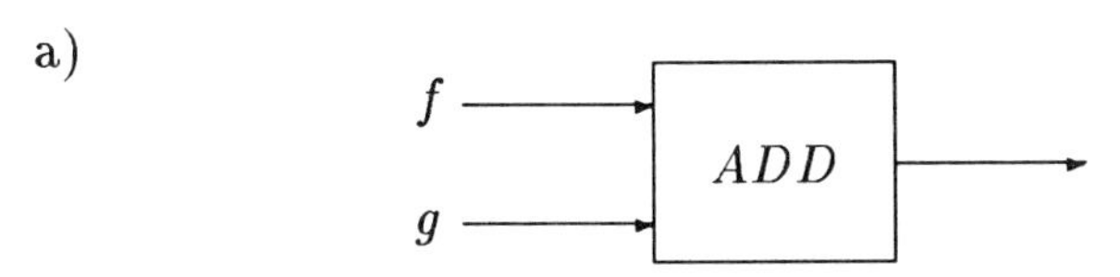

b)

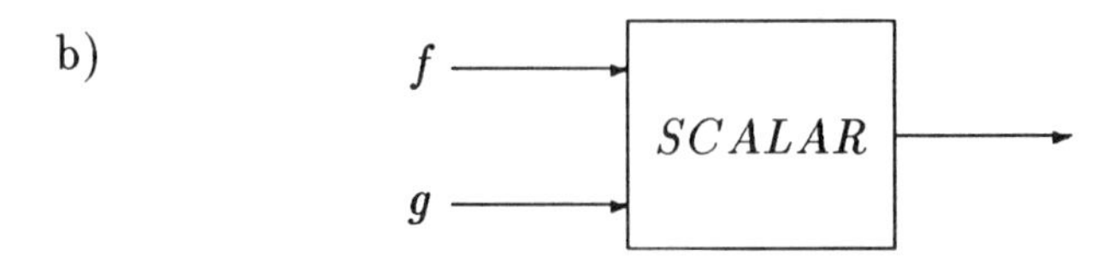

c)

g → $1/z$ →

d)

g → $NINETY$ →

3. For what class of digital signals f is it true that

$$\mathcal{Z}(f) = \mathcal{I}(\mathcal{Z}(f))$$

4. For what class of digital signals f is it true that

$$\mathcal{Z}(f) = -\mathcal{I}(\mathcal{Z}(f))$$

5. For f and g given in Exercise 1, find $f * g$ and verify the result by finding $\mathcal{Z}^{-1}(F(z,w)G(z,w))$.

6. Answer the same question as in Exercise 5 using

$$f = \begin{pmatrix} 1 & 0 & 0 & 0 & 0 \\ 3 & 4 & -1 & 0 & 2 \end{pmatrix}_{5,1}^{0}$$

and

$$g = \begin{pmatrix} 4 & 1 \\ 2 & 0 \end{pmatrix}_{0,0}^{0}$$

7. Consider analytic function $F(z,w) = 1/(1+zw)$. We can write

$$F(z,w) = 1 - zw + (zw)^2 - (zw)^3 + \cdots$$

for $|zw| < 1$ and therefore

$$f = \begin{pmatrix} \cdots & 0 & 0 & 0 & \boxed{1} \\ \cdots & 0 & 0 & -1 & 0 \\ \cdots & 0 & 1 & 0 & 0 \\ \cdots & -1 & 0 & 0 & 0 \\ & & & & \\ & & & & \end{pmatrix}^0$$

has $\mathcal{Z}$ transform $F(z, w)$. On the other hand, we can write

$$F(z, w) = \frac{1}{zw} \cdot \frac{1}{1 + 1/zw} = (zw)^{-1} - (zw)^{-2} + (zw)^{-3} - (zw)^{-4} + \cdots$$

for $|zw| > 1$ and

$$g = \begin{pmatrix} \cdot & \cdot & \cdot & \cdot & \\ \cdot & \cdot & \cdot & \cdot & \\ \cdot & \cdot & \cdot & \cdot & \\ 0 & 0 & 0 & 1 & \cdots \\ 0 & 0 & -1 & 0 & \cdots \\ 0 & 1 & 0 & 0 & \cdots \\ \boxed{0} & 0 & 0 & 0 & \cdots \end{pmatrix}^0$$

also has $\mathcal{Z}$ transform $F(z, w)$. Explain why the results are not a contradiction to the one-to-one correspondence between $\Re^{Z \times Z}$ and all $\mathcal{Z}$ transforms.

8. Find two distinct Laurent expansions for analytic function

$$F(z, w) = \frac{1}{1 - 3zw}$$

9. Find the Laurent expansion for $e^{1/zw}$. Where does the series converge?

10. Find Laurent expansion for $\cos(zw)$.

11. Can a two dimensional analytic function have isolated singularities?

12. How many different Laurent expansion are there for $(z+w)^{-1}$?

13. Find $COR(f,g)$ using $\mathcal{Z}$ transforms where

$$f = \begin{pmatrix} 1 & 0 \\ 2 & 3 \end{pmatrix}^{0}_{0,0}$$

$$g = \begin{pmatrix} 2 & 0 \\ 4 & 6 \end{pmatrix}^{0}_{2,3}$$

14. Find $COR(f,g)$ using $\mathcal{Z}$ transforms where

$$f = \begin{pmatrix} 1 & 0 & 0 & 0 \\ 2 & 4 & 1 & 3 \end{pmatrix}^{0}_{5,1}$$

$$g = \begin{pmatrix} 1 & -1 & 0 & 3 \\ -1 & 0 & 0 & 0 \end{pmatrix}^{0}_{2,0}$$

15. Using $\mathcal{Z}$ transforms to find transfer function if the input signal f and output g are

$$f = \begin{pmatrix} 2 & 0 & 1 \end{pmatrix}^{0}_{0,0}$$

$$g = \begin{pmatrix} 0 & 0 & 1 \\ 0 & 0 & 0 \\ 2 & 0 & 0 \end{pmatrix}^{0}_{0,0}$$

16. In Example 4.22 the input signal

$$f = \begin{pmatrix} 1 & -1 \end{pmatrix}^{0}_{0,0}$$

was given and the impulse function

$$h = \begin{pmatrix} \vdots & \vdots & \vdots & \\ 0 & 0 & 0 & \cdots \\ 0 & -1 & -1 & \cdots \\ \boxed{1} & 1 & 1 & \cdots \end{pmatrix}^{0}$$

was found. Convolve f with h to show that

$$g = \begin{pmatrix} 0 & -1 \\ 1 & 0 \end{pmatrix}_{0,1}^{0}$$

17. Repeat Exercise 16 using

$$h = \begin{pmatrix} \cdots & 1 & 1 & \boxed{1} \\ \cdots & -1 & -1 & 0 \\ \cdots & 0 & 0 & 0 \\ & \cdot & \cdot & \cdot \\ & \cdot & \cdot & \cdot \\ & \cdot & \cdot & \cdot \end{pmatrix}^{0}$$

5

Difference Equations

5.1 Function Equations

In this chapter we will study a very small, but important class of difference equations. Difference equations are a special type of function equation. A function equation involves one or more unknown digital signals (in $\Re^{Z\times Z}$) connected by the fundamental operations: $+$, $\cdot$, $\vee$, N, S, D, or macro operations. The objective of a function equation is to find, if it exists, a solution. Several cases arise. There may be no solutions, a single solution, or many solutions. Even when solutions exist, they are not easy to find.

Example 5.1 *Let*

$$f \cdot f = -1_{Z\times Z}$$

then there are no solutions in $\Re^{Z\times Z}$.

Example 5.2 *Consider the function equation*

$$f \cdot f = 1_{Z\times Z}$$

This equation, although very simple, has an infinite number of solutions. Indeed, at any point (n, m) *in* $\mathbf{Z} \times \mathbf{Z}$, $f(n, m)$ *can equal* 1 *or* -1. *Two solutions are*

$$f = \begin{pmatrix} & \cdot & \cdot & \cdot & \cdot & \cdot & \cdot & \\ & \cdot & \cdot & \cdot & \cdot & \cdot & \cdot & \\ & \cdot & \cdot & \cdot & \cdot & \cdot & \cdot & \\ \cdots & -1 & -1 & -1 & 1 & 1 & 1 & \cdots \\ \cdots & -1 & -1 & -1 & \boxed{1} & 1 & 1 & \cdots \\ \cdots & -1 & -1 & -1 & 1 & 1 & 1 & \cdots \\ & \cdot & \cdot & \cdot & \cdot & \cdot & \cdot & \\ & \cdot & \cdot & \cdot & \cdot & \cdot & \cdot & \\ & \cdot & \cdot & \cdot & \cdot & \cdot & \cdot & \end{pmatrix}$$

and

$$f = \begin{pmatrix} & \vdots & \vdots & \vdots & \vdots & \vdots & \vdots & \\ \cdots & 1 & -1 & 1 & -1 & 1 & -1 & \cdots \\ \cdots & 1 & -1 & 1 & \boxed{-1} & 1 & -1 & \cdots \\ \cdots & 1 & -1 & 1 & -1 & 1 & -1 & \cdots \\ & \vdots & \vdots & \vdots & \vdots & \vdots & \vdots & \end{pmatrix}$$

Example 5.3 *The function equation*

$$f \vee f_{1,0} = 0_{Z\times Z}$$

also has an infinite number of solutions. At any point (n, m)

$$f(n,m) \vee f(n-1,m) = 0$$

must hold. Thus, any signal in $\Re^{Z\times Z}$ *which is zero for at least one out of every pair of consecutive horizontal points and negative valued elsewhere is a solution. Two typical solutions are*

$$f = \begin{pmatrix} & \vdots & \vdots & \vdots & \vdots & \vdots & \vdots & & \\ \cdots & 0 & -1 & 0 & -1 & 0 & -1 & 0 & \cdots \\ \cdots & \boxed{0} & -1 & 0 & -1 & 0 & -1 & 0 & \cdots \\ & \vdots & \vdots & \vdots & \vdots & \vdots & \vdots & & \end{pmatrix}$$

and

$$f = \begin{pmatrix} & \vdots & \vdots & \vdots & \vdots & \vdots & \vdots & & \\ \cdots & 0 & -1 & 0 & -2 & 0 & -3 & 0 & \cdots \\ \cdots & \boxed{0} & -1 & 0 & -2 & 0 & -3 & 0 & \cdots \\ & \vdots & \vdots & \vdots & \vdots & \vdots & \vdots & & \end{pmatrix}$$

Example 5.4 *Suppose that the function equation*

$$f^+ + f^- = 0_{Z\times Z}$$

is given. In this case, there is a unique solution:

$$f = 0_{Z\times Z}$$

A difference equation is a function equation involving two or more translates of the unknown function. Accordingly, the function equation in Example 5.3 is a difference equation. It is a function equation involving $f_{0,0}$ and $f_{1,0}$. However, the difference equations which we will investigate herein will only involve the additional operations of $SCALAR$, ADD, and $MINUS$.

Example 5.5 *Several types of difference equations involving just the* $SCALAR$, ADD, *and* $MINUS$ *are:*

$$f_{3,2} + 4f_{1,4} = 5_{Z\times Z}$$

$$7f_{2,1} - 4f_{1,0} + 3f = 5$$

$$f - 3f_{1,3} = 0_{Z\times Z}$$

Systematic way of solving some difference equations similar to the ones given in the last example will be investigated in forthcoming sections.

5.2 Linear Space Invariant Difference Equations

As mentioned in the previous section, we will mainly be interested in difference equations involving only the operations of $SCALAR$, ADD, and $MINUS$. Such difference equations are often classified as linear space invariant. A typical equation, considered herein, of this form can be written as a finite sum involving lattice points from some designated set A.

$$\sum_{(i,j)\in A} h(i,j)g_{i,j} = f$$

where $h(i,j)$ are real numbers called the coefficients of the difference equation. The function f is a known digital signal, often called the driving or excitation function and the digital signal g is to be found. It is called the response signal.

Example 5.6 *The following difference equations*

$$g - g_{2,2} = 2_{Z\times Z}$$

$$g + 4g_{1,4} + 4g_{2,1} = u$$

$$g + 5g_{1,-1} + 4g_{2,3} = 3_{Z\times Z}$$

involve the index set of translations

$$A = \{(0,0),(2,2)\}$$

$$A = \{(0,0),(1,4),(2,1)\}$$

$$A = \{(0,0),(1,-1),(2,3)\}$$

respectively.

Example 5.7 *The next two difference equations*

$$g - g_{1,0} = 0_{Z\times Z}$$

$$g - 4g_{1,1} = 1_{Z\times Z}$$

involve the index set of translations

$$A = \{(0,0),(1,0)\}$$

and

$$A = \{(0,0),(1,1)\}$$

respectively. Moreover, in the first difference equation

$$h(0,0) = 1$$

$$h(1,0) = -1$$

and

$$f = 0_{Z\times Z}$$

As seen in the previous section, the solution, if one exists, to a difference equation need not be unique. When additional constraints are imposed the desired signal, sometimes a unique solution will result. Typical types of constraints involve support in some region, usually a quadrant or some finite set. Consider the next example where no constraints are given.

Example 5.8 *Suppose that an arbitrary solution is to be found for the equation*

$$g - g_{1,0} = \delta$$

Pointwise, this difference equation corresponds to

$$g(n,m) - g(n-1,m) = 0 \quad \textit{for all } m \neq 0$$

and

$$g(n,0) - g(n-1,0) = \begin{cases} 1 & n = 0 \\ 0 & \textit{otherwise} \end{cases}$$

The function equation can be written as

$$\begin{pmatrix} & \vdots & \vdots & \vdots & \\ \cdots & g(-1,1) - g(-2,1) & g(0,1) - g(-1,1) & g(1,1) - g(0,1) & \cdots \\ \cdots & g(-1,0) - g(-2,0) & \boxed{g(0,0) - g(-1,0)} & g(1,0) - g(0,0) & \cdots \\ \cdots & g(-1,-1) - g(-2,-1) & g(0,-1) - g(-1,-1) & g(1,-1) - g(0,-1) & \cdots \\ & \vdots & \vdots & \vdots & \end{pmatrix}$$

$$= \begin{pmatrix} & \vdots & \vdots & \vdots & \\ \cdots & 0 & 0 & 0 & \cdots \\ \cdots & 0 & \boxed{1} & 0 & \cdots \\ \cdots & 0 & 0 & 0 & \cdots \\ & \vdots & \vdots & \vdots & \end{pmatrix}$$

In any case, we see that for $m \neq 0$

$$g(n,m) = g(n-1,m)$$

is always true. So the digital signal as given by the bound matrix must have every row (except the $m = 0$ row) equal to a constant.

Additionally, for the row $m = 0$, there are several cases to consider. The first case is when $n < 0$, then we always have

$$g(n,0) - g(n-1,0) = 0$$

So in this case,

$$g(-1,0) = g(-2,0) = g(-3,0) = \cdots$$

holds true. This implies that for $n < 0$, and $m = 0$, g is also a constant. The second case is for $n = 0$, and $m = 0$, here,

$$g(0,0) - g(-1,0) = 1$$

Thus, if we use from case 1

$$g(-1,0) = a$$

then

$$g(0,0) = a + 1$$

The third and final case is for $n > 0$ and $m = 0$. Now

$$g(n,0) = g(n-1,0) = 0$$

and so we obtain

$$g(0,0) = g(1,0) = g(2,0) = \cdots$$

which must equal $a + 1$. Two typical solutions are

$$g = \begin{pmatrix} & \cdot & \cdot & \cdot & \cdot & \cdot & \\ & \cdot & \cdot & \cdot & \cdot & \cdot & \\ & \cdot & \cdot & \cdot & \cdot & \cdot & \\ \cdots & 2 & 2 & 2 & 2 & 2 & \cdots \\ \cdots & 1 & 1 & 1 & 1 & 1 & \cdots \\ \cdots & 0 & 0 & \boxed{1} & 1 & 1 & \cdots \\ \cdots & -1 & -1 & -1 & -1 & -1 & \cdots \\ \cdots & -2 & -2 & -2 & -2 & -2 & \cdots \\ & \cdot & \cdot & \cdot & \cdot & \cdot & \\ & \cdot & \cdot & \cdot & \cdot & \cdot & \\ & \cdot & \cdot & \cdot & \cdot & \cdot & \end{pmatrix}$$

and

$$g = \begin{pmatrix} 1 & 1 & 1 & \cdots \end{pmatrix}_{0,0}^{0}$$

We will "redo" the previous example. This time using first quadrant signals.

Example 5.9 *Suppose we want to find the solution of*

$$g - g_{1,0} = \delta$$

This time we will first assume that g is a first quadrant signal. Accordingly, g can be written

$$g = \begin{pmatrix} \cdot & \cdot & \\ \cdot & \cdot & \\ \cdot & \cdot & \\ g(0,1) & g(1,1) & \cdots \\ \boxed{g(0,0)} & g(1,0) & \cdots \end{pmatrix}^0$$

Referring to the in-depth discussion given in Example 5.8, we see that a unique solution exists in this case. It is

$$g = \begin{pmatrix} 1 & 1 & 1 & \cdots \end{pmatrix}^0_{0,0}$$

If instead we assume that g is a second quadrant signal then

$$g = \begin{pmatrix} & \cdot & \cdot \\ & \cdot & \cdot \\ & \cdot & \cdot \\ \cdots & g(-1,1) & g(0,1) \\ \cdots & g(-1,0) & \boxed{g(0,0)} \end{pmatrix}^0$$

Here the solution is given by

$$g = \begin{pmatrix} \cdots & -1 & -1 & \boxed{-1} \end{pmatrix}^0$$

Notice that if g is restricted to be a fourth quadrant signal then a similar solution to the one given by the first quadrant signal arises. Additionally, if g is restricted to be a third quadrant signal then the solution is similar to the second quadrant signal above.

All solutions can be checked by direct substitution into the defining equations. Indeed, equational identities must result from this substitution.

Example 5.10 *Assume that a first quadrant signal g is to be found which satisfies the difference equation:*

$$g - 2g_{1,1} = \delta$$

Here

$$g = \begin{pmatrix} \cdot & \cdot & \cdot & \\ \cdot & \cdot & \cdot & \\ \cdot & \cdot & \cdot & \\ g(0,2) & g(1,2) & g(2,2) & \cdots \\ g(0,1) & g(1,1) & g(2,1) & \cdots \\ \boxed{g(0,0)} & g(1,0) & g(2,0) & \cdots \end{pmatrix}^{0}$$

and so

$$g_{1,1} = \begin{pmatrix} \cdot & \cdot & \cdot & \\ \cdot & \cdot & \cdot & \\ \cdot & \cdot & \cdot & \\ 0 & g(0,1) & g(1,1) & \cdots \\ 0 & g(0,0) & g(1,0) & \cdots \\ \boxed{0} & 0 & 0 & \cdots \end{pmatrix}^{0}$$

$$g - 2g_{1,1} = \begin{pmatrix} \cdot & \cdot & \cdot & \\ \cdot & \cdot & \cdot & \\ \cdot & \cdot & \cdot & \\ g(0,2) & g(1,2) - 2g(0,1) & g(2,1) - 2g(1,1) & \cdots \\ g(0,1) & g(1,1) - 2g(0,0) & g(2,1) - 2g(1,0) & \cdots \\ \boxed{g(0,0)} & g(1,0) & g(2,0) & \cdots \end{pmatrix}^{0}$$

$$= \begin{pmatrix} \cdot & \cdot & \cdot & \\ \cdot & \cdot & \cdot & \\ \cdot & \cdot & \cdot & \\ 0 & 0 & 0 & \cdots \\ 0 & 0 & 0 & \cdots \\ \boxed{1} & 0 & 0 & \cdots \end{pmatrix}^{0}$$

So every entry in the bottom row of this bound matrix for g is zero except for

$$g(0,0) = 1$$

The same is true for the left most column of g; every entry there is zero except for $g(0,0)$. Entries along the 45° major diagonal in the bound matrix for $g - 2g_{1,1}$ gives

$$\begin{aligned}
&g(0,0) = 1 \\
&g(1,1) - 2g(0,0) = 0 \text{ which implies } g(1,1) = 2 \\
&g(2,2) - 2g(1,1) = 0 \text{ which implies } g(2,2) = 2^2 \\
&\cdot \\
&\cdot \\
&\cdot \\
&g(n,n) - 2g(n-1,n-1) = 0 \text{ which implies } g(n,n) = 2^n
\end{aligned}$$

Similarly, entries along 45° lines not on the major diagonal provide pointwise recursive equations which when solved give value 0. The solution is

$$g = \begin{pmatrix} \cdot & \cdot & \cdot & \\ \cdot & \cdot & \cdot & \\ \cdot & \cdot & \cdot & \\ 0 & 0 & 2^2 & \cdots \\ 0 & 2^1 & 0 & \cdots \\ \boxed{1} & 0 & 0 & \cdots \end{pmatrix}^0$$

Notice that

$$g_{1,1} = \begin{pmatrix} \cdot & \cdot & \cdot & \\ \cdot & \cdot & \cdot & \\ \cdot & \cdot & \cdot & \\ 0 & 0 & 2^1 & \cdots \\ 0 & 1 & 0 & \cdots \\ \boxed{0} & 0 & 0 & \cdots \end{pmatrix}^0$$

and

$$-2g_{1,1} = \begin{pmatrix} \cdot & \cdot & \cdot & \\ \cdot & \cdot & \cdot & \\ \cdot & \cdot & \cdot & \\ 0 & 0 & -2^2 & \cdots \\ 0 & -2^1 & 0 & \cdots \\ \boxed{0} & 0 & 0 & \cdots \end{pmatrix}^0$$

Thus,

$$g - 2g_{1,1} = \delta$$

holds true, thereby verifying that g is the solution to this difference equation.

5.3 Difference Equations Involving One Unit Translation

In this section we will present systematic procedures for solving difference equations of the form

$$g - ag_{n,m} = f$$

Here, a is a nonzero real number and both m and n equal 0 or 1, with both n and m not equal to zero. Moreover, f is known and the objective is to find a first quadrant signal which results in an identity when substituted into this equation.

While only the first quadrant signals are considered herein, the same procedures can be employed for signals with support on the other regions. To begin, we will consider the simple difference equation

$$g - ag_{1,0} = 0_{Z\times Z}$$

Using bound matrices this is equivalent to

$$\begin{pmatrix} \cdot & \cdot & \cdot & \\ \cdot & \cdot & \cdot & \\ \cdot & \cdot & \cdot & \\ g(0,2) & g(1,2)-ag(0,2) & g(2,2)-ag(1,2) & \cdots \\ g(0,1) & g(1,1)-ag(0,1) & g(2,1)-ag(1,1) & \cdots \\ \boxed{g(0,0)} & g(1,0)-ag(0,0) & g(2,0)-ag(1,0) & \cdots \end{pmatrix}^{0} \tag{5.1}$$

$$= \begin{pmatrix} \cdot & \cdot & \cdot & \\ \cdot & \cdot & \cdot & \\ \cdot & \cdot & \cdot & \\ 0 & 0 & 0 & \cdots \\ 0 & 0 & 0 & \cdots \\ \boxed{0} & 0 & 0 & \cdots \end{pmatrix}^{0}$$

The entries in the first column in matrix 5.1 above must be all zero. Since g only has support in the first quadrant, it follows that

$$g(0,k) = 0 \quad \text{for } k = 0, 1, 2, \ldots$$

Substituting these values in the second column of matrix 5.1 above gives entries of the form

$$g(1,k) = ag(0,k) = 0$$

Accordingly, all entries

$$g(1,k) = 0$$

Similarly, all entries

$$g(2,k) = 0$$

and so the solution is

$$g = 0_{Z \times Z}$$

An analogous situation arises if we want to find the first quadrant signal g such that

$$g - ag_{0,1} = 0_{Z \times Z}$$

Indeed, the unique solution again is

$$g = 0_{Z \times Z}$$

In fact, using n and m as given above, the only first quadrant solution to

$$g - ag_{n,m} = 0_{Z \times Z}$$

is

$$g = 0_{Z \times Z}$$

When we have a difference equation, as above, with

$$f = 0_{Z \times Z}$$

it is called *homogeneous.* Otherwise, it is called *non-homogeneous.* In all cases, f must have support in the first quadrant since the desired solution has support in this quadrant.

Consider

$$g - ag_{1,0} = f$$

In bound matrix form we have:

$$\begin{pmatrix} \cdot & \cdot & \cdot & \\ \cdot & \cdot & \cdot & \\ \cdot & \cdot & \cdot & \\ g(0,2) & g(1,2) - ag(0,2) & g(2,2) - ag(1,2) & \cdots \\ g(0,1) & g(1,1) - ag(0,1) & g(2,1) - ag(1,1) & \cdots \\ \boxed{g(0,0)} & g(1,0) - ag(0,0) & g(2,0) - ag(1,0) & \cdots \end{pmatrix}^{0}$$

$$= \begin{pmatrix} \cdot & \cdot & \cdot & \\ \cdot & \cdot & \cdot & \\ \cdot & \cdot & \cdot & \\ f(0,2) & f(1,2) & f(2,2) & \cdots \\ f(0,1) & f(1,1) & f(2,1) & \cdots \\ \boxed{f(0,0)} & f(1,0) & f(2,0) & \cdots \end{pmatrix}^0$$

The entries in the left most column of both bound matrices are equal. Accordingly,

$$g(0,k) = f(0,k) \quad k = 0,1,2,...$$

Similarly, the entries in the second column in both matrices are equal. This gives

$$g(1,k) - ag(0,k) = f(1,k) \quad k = 0,1,2,...$$

which is equivalent to

$$g(1,k) = af(0,k) + f(1,k) \quad k = 0,1,2,...$$

Equating entries in the third column gives

$$g(2,k) - ag(1,k) = f(2,k) \quad k = 0,1,2,...$$

which is equivalent to

$$g(2,k) = a^2 f(0,k) + af(1,k) + f(2,k)$$

and so on. In general, we obtain

$$g(n,k) = \sum_{i=0}^{n} a^i f(n-i,k) \quad \text{for all } k = 0,1,2,...$$

Observe that this solution can be expressed as a convolution involving the signals

$$h = \begin{pmatrix} \boxed{1} & a & a^2 & \cdots \end{pmatrix}_{0,0}^{0}$$

and f. Thus, we have

$$g = f * h$$

Example 5.11 *Find the first quadrant signal g such that*

$$g - 2g_{1,0} = \begin{pmatrix} 3 & 0 \\ 2 & 4 \\ 1 & 0 \end{pmatrix}_{0,2}^{0}$$

We will form the convolution of

$$h = \begin{pmatrix} \boxed{1} & 2 & 4 & \cdots \end{pmatrix}^{0}$$

with f using the parallel algorithm. First, we find that

$$coz(f) = \{(0,2),\ (0,1),\ (0,0),\ (1,1)\}$$

Next, form the scalar translation connections

$$3h_{0,2} = \begin{pmatrix} 3 & 6 & 12 & \cdots \\ 0 & 0 & 0 & \cdots \\ 0 & 0 & 0 & \cdots \end{pmatrix}_{0,2}^{0}$$

$$2h_{0,1} = \begin{pmatrix} 0 & 0 & 0 & \cdots \\ 2 & 4 & 0 & \cdots \\ 0 & 0 & 0 & \cdots \end{pmatrix}_{0,2}^{0}$$

$$h_{0,0} = \begin{pmatrix} 0 & 0 & 0 & \cdots \\ 0 & 0 & 0 & \cdots \\ 1 & 2 & 4 & \cdots \end{pmatrix}_{0,2}^{0}$$

$$4h_{1,1} = \begin{pmatrix} 0 & 0 & 0 & \cdots \\ 0 & 4 & 8 & \cdots \\ 0 & 0 & 0 & \cdots \end{pmatrix}_{0,2}^{0}$$

Next, adding all the above terms involving h gives the solution:

$$g = \begin{pmatrix} 3 & 6 & 12 & 24 & \cdots \\ 2 & 8 & 16 & 32 & \cdots \\ 1 & 2 & 4 & 8 & \cdots \end{pmatrix}_{0,2}^{0}$$

Notice that

$$-2g_{1,0} = \begin{pmatrix} 0 & -6 & -12 & -24 & \cdots \\ 0 & -4 & -16 & -32 & \cdots \\ 0 & -2 & -4 & -8 & \cdots \end{pmatrix}_{0,2}^{0}$$

and so

$$g - 2g_{1,0} = \begin{pmatrix} 3 & 0 & 0 & 0 & \cdots \\ 2 & 4 & 0 & 0 & \cdots \\ 1 & 0 & 0 & 0 & \cdots \end{pmatrix}_{0,2}^{0} = f$$

If we now consider the equation

$$g - ag_{1,1} = f$$

where a is a nonzero real number and f is a known first quadrant signal, and g is a first quadrant signal to be found, then the solution is again given by the convolution formula:

$$g = f * \begin{pmatrix} \cdot & \cdot & \cdot & \\ \cdot & \cdot & \cdot & \\ \cdot & \cdot & \cdot & \\ 0 & 0 & a^2 & \cdots \\ 0 & a & 0 & \cdots \\ 1 & 0 & 0 & \cdots \end{pmatrix}^{0}$$

Example 5.12 *Solve*

$$g + 2g_{1,1} = \begin{pmatrix} -3 & 1 \\ 2 & 0 \\ 0 & -1 \end{pmatrix}_{0,2}^{0} = f$$

The solution will be given by the convolution formula

$$g = f * h$$

where

$$h = \begin{pmatrix} \cdot & \cdot & \cdot & \cdot & \\ \cdot & \cdot & \cdot & \cdot & \\ \cdot & \cdot & \cdot & \cdot & \\ 0 & 0 & 0 & -8 & \cdots \\ 0 & 0 & 4 & 0 & \cdots \\ 0 & -2 & 0 & 0 & \cdots \\ \boxed{1} & 0 & 0 & 0 & \cdots \end{pmatrix}^{0}$$

Again, we will use the parallel convolution method. First we find

$$coz(f) = \{(0,2), (0,1), (1,2), (1,0)\}$$

and forming the scalar translation connections involving h gives

$$-3h_{0,2} = \begin{pmatrix} \cdot & \cdot & \cdot & \cdot & \cdot \\ \cdot & \cdot & \cdot & \cdot & \cdot \\ \cdot & \cdot & \cdot & \cdot & \cdot \\ 0 & 0 & -12 & 0 & \cdots \\ 0 & 6 & 0 & 0 & \cdots \\ -3 & 0 & 0 & 0 & \cdots \\ 0 & 0 & 0 & 0 & \cdots \\ \boxed{0} & 0 & 0 & 0 & \cdots \end{pmatrix}^0$$

$$2h_{0,1} = \begin{pmatrix} \cdot & \cdot & \cdot & \cdot & \cdot \\ \cdot & \cdot & \cdot & \cdot & \cdot \\ \cdot & \cdot & \cdot & \cdot & \cdot \\ 0 & 0 & 8 & 0 & \cdots \\ 0 & -4 & 0 & 0 & \cdots \\ 2 & 0 & 0 & 0 & \cdots \\ \boxed{0} & 0 & 0 & 0 & \cdots \end{pmatrix}^0$$

$$1h_{1,2} = \begin{pmatrix} \cdot & \cdot & \cdot & \cdot & \cdot \\ \cdot & \cdot & \cdot & \cdot & \cdot \\ \cdot & \cdot & \cdot & \cdot & \cdot \\ 0 & 0 & 0 & 4 & \cdots \\ 0 & 0 & -2 & 0 & \cdots \\ 0 & 1 & 0 & 0 & \cdots \\ 0 & 0 & 0 & 0 & \cdots \\ \boxed{0} & 0 & 0 & 0 & \cdots \end{pmatrix}^0$$

$$-1h_{0,1} = \begin{pmatrix} \cdot & \cdot & \cdot & \cdot & \cdot \\ \cdot & \cdot & \cdot & \cdot & \cdot \\ \cdot & \cdot & \cdot & \cdot & \cdot \\ 0 & 0 & 0 & 0 & \cdots \\ 0 & 0 & 0 & 0 & \cdots \\ 0 & 0 & 0 & -4 & \cdots \\ 0 & 0 & 2 & 0 & \cdots \\ \boxed{0} & -1 & 0 & 0 & \cdots \end{pmatrix}^0$$

Adding these terms gives the solution

$$g = \begin{pmatrix} \cdot & \cdot & \cdot & \cdot & \\ \cdot & \cdot & \cdot & \cdot & \\ \cdot & \cdot & \cdot & \cdot & \\ 0 & 0 & -12 & -12 & \cdots \\ 0 & 6 & 6 & 0 & \cdots \\ -3 & -3 & 0 & -4 & \cdots \\ 2 & 0 & 2 & 0 & \cdots \\ \boxed{0} & -1 & 0 & 0 & \cdots \end{pmatrix}^0$$

Notice that, from Example 5.12, we have

$$2g_{1,1} = \begin{pmatrix} \cdot & \cdot & \cdot & \cdot & \\ \cdot & \cdot & \cdot & \cdot & \\ \cdot & \cdot & \cdot & \cdot & \\ 0 & 0 & 12 & 12 & \cdots \\ 0 & -6 & -6 & 0 & \cdots \\ 0 & 4 & 0 & 4 & \cdots \\ 0 & 0 & -2 & 0 & \cdots \\ \boxed{0} & 0 & 0 & 0 & \cdots \end{pmatrix}^0$$

and

$$g + 2g_{1,1} = \begin{pmatrix} \cdot & \cdot & \cdot & \cdot & \\ \cdot & \cdot & \cdot & \cdot & \\ \cdot & \cdot & \cdot & \cdot & \\ 0 & 0 & 0 & 0 & \cdots \\ 0 & 0 & 0 & 0 & \cdots \\ -3 & 1 & 0 & 0 & \cdots \\ 2 & 0 & 0 & 0 & \cdots \\ \boxed{0} & -1 & 0 & 0 & \cdots \end{pmatrix}^0$$

We can extend our method for n and $m = -1$, 0 and 1 (again both n, $m \neq 0$). This time, consider the difference equation

$$g - ag_{-1,-1} = f$$

where a is a nonzero real value.

We know that the solution of

$$g - ag_{1,1} = f$$

is

$$g = f * \begin{pmatrix} \cdot & \cdot & \cdot & \\ \cdot & \cdot & \cdot & \\ \cdot & \cdot & \cdot & \\ 0 & 0 & a^2 & \cdots \\ 0 & a & 0 & \cdots \\ \boxed{1} & 0 & 0 & \cdots \end{pmatrix}^0$$

Therefore, taking the translation of both sides of the equation below one unit up and one unit to the right

$$g - ag_{-1,-1} = f$$

gives

$$g_{1,1} - ag = f_{1,1}$$

Then, dividing by $-a$ (since $a \neq 0$) gives the equation

$$g - \left(\frac{1}{a}\right) g_{1,1} = \left(-\frac{1}{a}\right) f_{1,1}$$

The first quadrant solution of this equation is

$$g = \left(-\frac{1}{a}\right) f_{1,1} * h$$

and is given by the block diagram

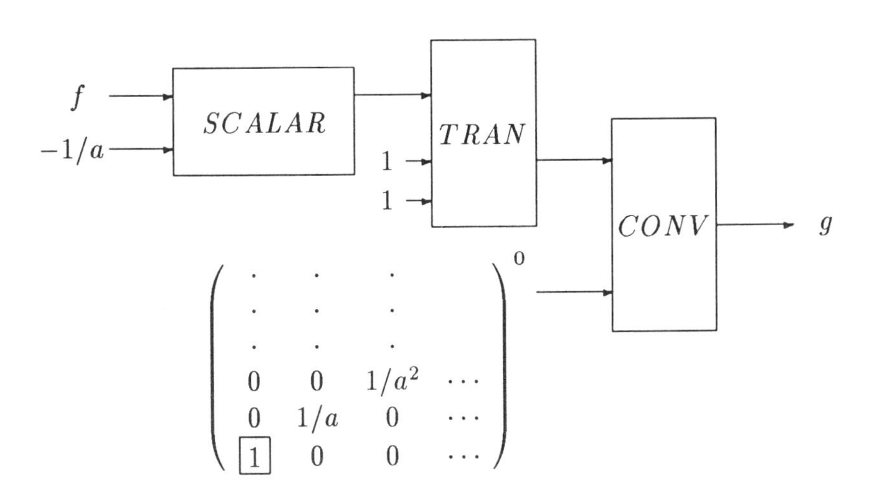

where

$$h = \begin{pmatrix} \cdot & \cdot & \cdot & \\ \cdot & \cdot & \cdot & \\ \cdot & \cdot & \cdot & \\ 0 & 0 & 1/a^2 & \cdots \\ 0 & 1/a & 0 & \cdots \\ \boxed{1} & 0 & 0 & \cdots \end{pmatrix}^0$$

Example 5.13 *Find a first quadrant solution g to*

$$g - 2g_{-1,-1} = \begin{pmatrix} 2 & 2 \\ 4 & -4 \end{pmatrix}^0_{0,1}$$

Notice that $a = 2$ here and

$$f_{1,1} = \begin{pmatrix} 0 & 2 & 2 \\ 0 & 4 & -4 \\ 0 & 0 & 0 \end{pmatrix}^0_{0,2}$$

In this case,

$$h = \begin{pmatrix} \cdot & \cdot & \cdot & \\ \cdot & \cdot & \cdot & \\ \cdot & \cdot & \cdot & \\ 0 & 0 & 1/4 & \cdots \\ 0 & 1/2 & 0 & \cdots \\ \boxed{1} & 0 & 0 & \cdots \end{pmatrix}^0$$

The co-zero set for $f_{1,1}$ is

$$coz(f) = \{(1,2), (1,1), (2,2), (2,1)\}$$

Next, we form the scalar translation connection involving h. So:

$$2h_{1,2} = \begin{pmatrix} \cdot & \cdot & \cdot & \cdot & \cdot & \\ \cdot & \cdot & \cdot & \cdot & \cdot & \\ \cdot & \cdot & \cdot & \cdot & \cdot & \\ 0 & 0 & 0 & 1/2 & 0 & \cdots \\ 0 & 0 & 1 & 0 & 0 & \cdots \\ 0 & 2 & 0 & 0 & 0 & \cdots \\ 0 & 0 & 0 & 0 & 0 & \cdots \\ \boxed{0} & 0 & 0 & 0 & 0 & \cdots \end{pmatrix}^0$$

$$4h_{1,1} = \begin{pmatrix} \cdot & \cdot & \cdot & \cdot & \cdot & \\ \cdot & \cdot & \cdot & \cdot & \cdot & \\ \cdot & \cdot & \cdot & \cdot & \cdot & \\ 0 & 0 & 0 & 0 & 1/2 & \cdots \\ 0 & 0 & 0 & 1 & 0 & \cdots \\ 0 & 0 & 2 & 0 & 0 & \cdots \\ 0 & 4 & 0 & 0 & 0 & \cdots \\ \boxed{0} & 0 & 0 & 0 & 0 & \cdots \end{pmatrix}^0$$

$$2h_{2,2} = \begin{pmatrix} \cdot & \cdot & \cdot & \cdot & \cdot & \\ \cdot & \cdot & \cdot & \cdot & \cdot & \\ \cdot & \cdot & \cdot & \cdot & \cdot & \\ 0 & 0 & 0 & 0 & 1/2 & \cdots \\ 0 & 0 & 0 & 1 & 0 & \cdots \\ 0 & 0 & 2 & 0 & 0 & \cdots \\ 0 & 0 & 0 & 0 & 0 & \cdots \\ \boxed{0} & 0 & 0 & 0 & 0 & \cdots \end{pmatrix}^0$$

$$-4h_{2,1} = \begin{pmatrix} \cdot & \cdot & \cdot & \cdot & \cdot & \\ \cdot & \cdot & \cdot & \cdot & \cdot & \\ \cdot & \cdot & \cdot & \cdot & \cdot & \\ 0 & 0 & 0 & 0 & 0 & \cdots \\ 0 & 0 & 0 & 0 & -1 & \cdots \\ 0 & 0 & 0 & -2 & 0 & \cdots \\ 0 & 0 & -4 & 0 & 0 & \cdots \\ \boxed{0} & 0 & 0 & 0 & 0 & \cdots \end{pmatrix}^0$$

Adding the last four terms together and multiplying by $-\frac{1}{2}$ gives the solution:

$$g = \left(-\frac{1}{2}\right) \begin{pmatrix} \cdot & \cdot & \cdot & \cdot & \cdot & \\ \cdot & \cdot & \cdot & \cdot & \cdot & \\ \cdot & \cdot & \cdot & \cdot & \cdot & \\ 0 & 0 & 0 & 1/2 & 1 & \cdots \\ 0 & 0 & 1 & 2 & -1 & \cdots \\ 0 & 2 & 4 & -2 & 0 & \cdots \\ 0 & 4 & -4 & 0 & 0 & \cdots \\ \boxed{0} & 0 & 0 & 0 & 0 & \cdots \end{pmatrix}^0$$

If we form

$$-2g_{-1,-1} = \begin{pmatrix} \cdot & \cdot & \cdot & \cdot & \cdot & \\ \cdot & \cdot & \cdot & \cdot & \cdot & \\ \cdot & \cdot & \cdot & \cdot & \cdot & \\ 0 & 0 & 0 & -1/4 & -1/2 & \cdots \\ 0 & 0 & 1/2 & 1 & 0 & \cdots \\ 0 & 1 & 2 & -1 & 0 & \cdots \\ 2 & 4 & -2 & 0 & 0 & \cdots \\ \boxed{4} & -4 & 0 & 0 & 0 & \cdots \end{pmatrix}^0$$

then

$$g - 2g_{-1,-1} = \begin{pmatrix} \cdot & \cdot & \cdot & \cdot & \cdot & \\ \cdot & \cdot & \cdot & \cdot & \cdot & \\ \cdot & \cdot & \cdot & \cdot & \cdot & \\ 0 & 0 & 0 & 0 & 0 & \cdots \\ 0 & 0 & 0 & 0 & 0 & \cdots \\ 2 & 2 & 0 & 0 & 0 & \cdots \\ \boxed{4} & -4 & 0 & 0 & 0 & \cdots \end{pmatrix}^0 = f$$

Other difference equations involving one unit translation can be solved in exactly similar ways to the ones given above.

5.4 $\mathcal{Z}$ Transforms for Solving Difference Equations

In the previous section we solved several difference equations of the form

$$g - ag_{m,n} = f$$

where m and $n = -1$, 0 or 1. Observe that this equation can be illustrated using block diagrams as

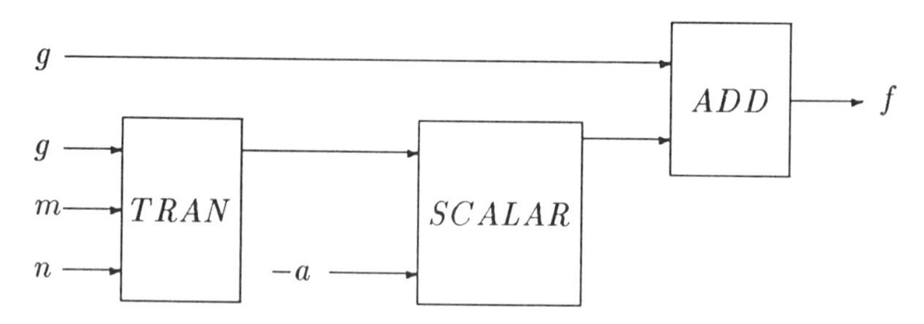

We immediately see that this is a convolution type representation. It involves the unknown signal g and a signal q called the coefficient signal.

Example 5.14 *The difference equation*

$$g - 2g_{1,1} = f$$

can be written as the convolution

$$g * \begin{pmatrix} 0 & -2 \\ 1 & 0 \end{pmatrix}_{0,1}^{0} = f$$

The above observation can be generalized to arbitrary linear space invariant difference equations.

Example 5.15 *Notice that the difference equation*

$$g - 3g_{1,0} + 2g_{1,1} + 4g_{2,3} = f$$

can be written as the convolution

$$g * \begin{pmatrix} 0 & 0 & 4 \\ 0 & 0 & 0 \\ 0 & 2 & 0 \\ 1 & -3 & 0 \end{pmatrix}_{0,3}^{0} = f$$

This observation provides the key to the $\mathcal{Z}$ transform method for solving difference equations of the above type.

We will let the $\mathcal{Z}$ transform of the coefficient signal be $Q(z,w)$. Thus, for instance, in Example 5.15 notice that

$$Q(z,w) = 1 - 3z^{-1} + 2z^{-1}w^{-1} + 4z^{-2}w^{-3}$$

Also let

$$\mathcal{Z}(f) = F(z,w) \quad \text{and} \quad \mathcal{Z}(g) = G(z,w)$$

Since the $\mathcal{Z}$ transform of the convolution of two signals is the product of their $\mathcal{Z}$ transforms, we obtain the all important equation

$$G(z,w)Q(z,w) = F(z,w)$$

Further manipulations in the transformed world involving indeterminates z and w gives

$$G(z,w) = \frac{1}{Q(z,w)} F(z,w)$$

Accordingly, if we divide the $\mathcal{Z}$ transform of the input signal by $Q(z,w)$ we will obtain the $\mathcal{Z}$ transform of the response signal. Finally, if we take an inverse $\mathcal{Z}$ transform of this quantity then the output signal will be found. Symbolically, we have

$$g = \mathcal{Z}^{-1}\left(\frac{\mathcal{Z}(f)}{\mathcal{Z}(q)}\right)$$

Notice that the transfer function, as described in Section 4.7 is given by

$$H(z,w) = \frac{1}{Q(z,w)}$$

Example 5.16 *Suppose that we wish to use $\mathcal{Z}$ transforms to solve the difference equation*

$$g + 2g_{1,1} = \begin{pmatrix} -3 & 1 \\ 2 & 0 \\ 0 & -1 \end{pmatrix}_{0,2}^{0} = f$$

As previously noted, this is equivalent to the convolution

$$g * \begin{pmatrix} 0 & 2 \\ 1 & 0 \end{pmatrix}_{0,1}^{0} = f$$

Consequently, taking $\mathcal{Z}$ transforms gives

$$G(z,w)(1 + 2z^{-1}w^{-1}) = -3w^{-2} + 2w^{-1} + z^{-1}w^{-2} - 1z^{-1}$$

So the $\mathcal{Z}$ transform of the solution is

$$G(z,w) = \frac{-3w^{-2} + 2w^{-1} + z^{-1}w^{-2} - 1z^{-1}}{1 + 2z^{-1}w^{-1}}$$

Expanding

$$\frac{1}{1 + 2z^{-1}w^{-1}}$$

into a Laurent expansion valid for

$$| 2z^{-1}w^{-1} |< 1$$

gives

$$1 - 2z^{-1}w^{-1} + 4z^{-2}w^{-2} - 8z^{-3}w^{-3} + \cdots$$

Multiplying this expansion by $\mathcal{Z}(f)$ *gives*

$$\begin{aligned} G(z,w) = \; & -3w^{-2} + 2w^{-1} + z^{-1}w^{-2} - z^{-1} \\ & -2z^{-1}w^{-1}(-3w^{-2} + 2w^{-1} + z^{-1}w^{-2} - z^{-1}) + \cdots \end{aligned}$$

Simplifying by combining terms and taking the inverse $\mathcal{Z}$ *transform gives*

$$g = \begin{pmatrix} \cdot & \cdot & \cdot & \cdot & \\ \cdot & \cdot & \cdot & \cdot & \\ \cdot & \cdot & \cdot & \cdot & \\ 0 & 6 & 6 & 0 & \cdots \\ -3 & -3 & 0 & -4 & \cdots \\ 2 & 0 & 2 & 0 & \cdots \\ \boxed{0} & -1 & 0 & 0 & \cdots \end{pmatrix}^{0}$$

We should mention that the solution obtained in the previous example is not unique. That solution has support in the first quadrant. There may be solutions with support on other regions. The reason for the non unique solution is related to the Laurent expansion. When it is found, there may be several expansions for a given region of convergence and each resulting in a different signal.

Example 5.17 *As in Example* 5.16, *again consider the difference equation*

$$g + 2g_{1,1} = \begin{pmatrix} -3 & 1 \\ 2 & 0 \\ 0 & -1 \end{pmatrix}_{0,2}^{0} = f$$

This time write $G(z,w)$ *in the equivalent, but different form*

$$G(z,w) = \frac{(1/2)zw[-3w^{-2} + 2w^{-1} + z^{-1}w^{-2} - z^{-1}]}{1 + (zw/2)}$$

$$= (1/2)[-3zw^{-1} + 2z + w^{-1} - w]\frac{1}{1 + (zw/2)}$$

Now expand

$$\frac{1}{1+(zw/2)}$$

in a Laurent expansion valid for

$$|\, zw/2 \,| < 1$$

gives

$$1-(zw/2)+(z^2w^2/4)-\cdots$$

Multiplying gives

$$G(z,w) = \left(\frac{1}{2}\right)\left[-3zw^{-1}+2z+w^{-1}-w\right]$$
$$-\left(\frac{1}{4}\right)\left[-3z^2+2z^2w+z-zw^2\right]$$
$$+\left(\frac{1}{8}\right)\left[-3z^3w+2z^3w^2+z^2w-z^2w^3\right]$$
$$-\left(\frac{1}{16}\right)\left[-3z^4w^2+2z^4w^3+z^3w^2-z^3w^4\right]+\cdots$$

Taking the inverse $\mathcal{Z}$ transform gives the other solution

$$g = \begin{pmatrix} \cdots & 0 & 0 & -3/2 & 1/2 \\ \cdots & 0 & 3/4 & 3/4 & \boxed{0} \\ \cdots & -3/8 & -3/8 & 0 & -1/2 \\ & \cdot & \cdot & \cdot & \cdot \\ & \cdot & \cdot & \cdot & \cdot \\ & \cdot & \cdot & \cdot & \cdot \end{pmatrix}^{0}$$

Notice that

$$2g_{1,1} = \begin{pmatrix} \cdots & 0 & 0 & -3 & 1 \\ \cdots & 0 & 3/2 & 3/2 & 0 \\ \cdots & -3/4 & -3/4 & \boxed{0} & -1 \\ & \cdot & \cdot & \cdot & \cdot \\ & \cdot & \cdot & \cdot & \cdot \\ & \cdot & \cdot & \cdot & \cdot \end{pmatrix}^{0}$$

Substituting in the equation yields

$$g + 2g_{1,1} = \begin{pmatrix} \cdots & 0 & 0 & -3 & 1 \\ \cdots & 0 & 0 & 2 & 0 \\ \cdots & 0 & 0 & \boxed{0} & -1 \\ & 0 & 0 & 0 & 0 \\ & \cdot & \cdot & \cdot & \cdot \\ & \cdot & \cdot & \cdot & \cdot \\ & \cdot & \cdot & \cdot & \cdot \end{pmatrix}^{0} = f$$

More general difference equations can be solved using $\mathcal{Z}$ transforms than those illustrated above. Indeed, linear difference equations with constant coefficients involving numerous translations of one unit or more can be solved. Again, we find the output transform $G(z, w)$ using

$$G(z, w) = \frac{F(z, w)}{Q(z, w)}$$

All that need be done is to divide the denominator transform Q into the numerator transform F and then take an inverse $\mathcal{Z}$ transform of the quotient. An additional example should make the technique more clear. Again, the region of convergence may govern the number and nature of the solution.

Example 5.18 *Find the signal g such that:*

$$g - g_{1,0} - g_{0,1} + g_{1,1} = \delta$$

In terms of convolution we have

$$g * \begin{pmatrix} 1 & 1 \\ 1 & -1 \end{pmatrix}_{0,1}^{0} = (1)_{0,0}^{0}$$

Taking the $\mathcal{Z}$ transform gives

$$G(z, w)[1 - w^{-1} - z^{-1} + z^{-1}w^{-1}] = 1$$

writing

$$G(z, w) = \frac{F(z, w)}{Q(z, w)}$$

gives

$$G(z, w) = \frac{1}{1 - w^{-1} - z^{-1} + z^{-1}w^{-1}}$$

We can actually divide "Long hand." However, due to the simplicity of this problem, we can write

$$G(z,w) = \frac{1}{1-w^{-1}}\frac{1}{1-z^{-1}}$$

$$= (1 + w^{-1} + w^{-2} + \cdots)(1 + z^{-1} + z^{-2} + \cdots)$$
$$= 1 + w^{-1} + z^{-1} + z^{-1}w^{-1} + w^{-2} + z^{-2}$$
$$+ z^{-2}w^{-1} + z^{-1}w^{-2} + z^{-2}w^{-2} + \cdots$$

for

$$|z^{-1}| < 1 \quad \text{and} \quad |w^{-1}| < 1$$

In any case, taking the inverse $\mathcal{Z}$ transform gives

$$g = \begin{pmatrix} \cdot & \cdot & \cdot & \\ \cdot & \cdot & \cdot & \\ \cdot & \cdot & \cdot & \\ 1 & 1 & 1 & \cdots \\ 1 & 1 & 1 & \cdots \\ \boxed{1} & 1 & 1 & \cdots \end{pmatrix}^{0}$$

This can be verified by forming

$$g - g_{1,0} - g_{0,1} + g_{1,1}$$

and noting that the result is δ. Indeed, observe that

$$g + g_{1,1} = \begin{pmatrix} \cdot & \cdot & \cdot & \\ \cdot & \cdot & \cdot & \\ \cdot & \cdot & \cdot & \\ 1 & 2 & 2 & \cdots \\ 1 & 2 & 2 & \cdots \\ \boxed{1} & 1 & 1 & \cdots \end{pmatrix}^{0}$$

and

$$-(g_{1,0} + g_{0,1}) = -\begin{pmatrix} \cdot & \cdot & \cdot & \\ \cdot & \cdot & \cdot & \\ \cdot & \cdot & \cdot & \\ 1 & 2 & 2 & \cdots \\ 1 & 2 & 2 & \cdots \\ \boxed{0} & 1 & 1 & \cdots \end{pmatrix}^{0}$$

Thus, adding these last two expressions together gives δ.

5.5 Exercises

1. Find all digital solutions to the function equation

$$f \wedge f = 0_{Z\times Z}$$

Is this equation a difference equation?

2. Find all solutions to the functional equation which is given below using block diagrams. Specifically, find all digital signals f, such that the output of the following block diagram is $0_{Z\times Z}$.

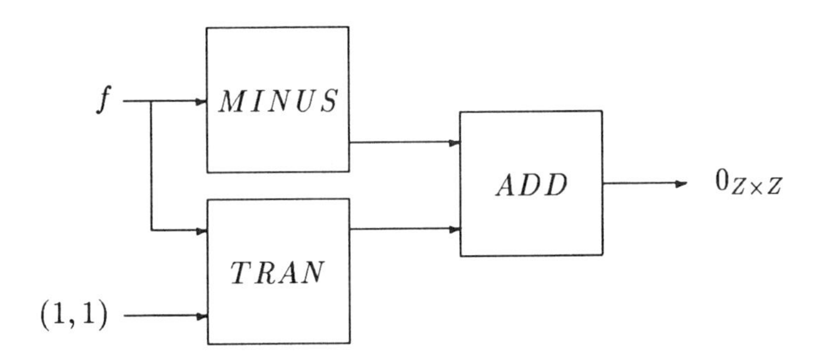

Is this a difference equation?

3. Find all solutions f to the following difference equation

$$f - f_{1,0} = 0_{Z\times Z}$$

4. Which of the following difference equations have at least one solution? Find the solution to each of the difference equations which have solutions.

a) $f^{+} \vee \mid f \mid = -1_{Z\times Z}$

b) $f^{+} - f^{-} = -1_{Z\times Z}$

c) $f \cdot f_{3,0} = 1_{Z\times Z}$

5. Find a first quadrant response signal g to the first order difference equation given by the block diagram whose output is the

impulse signal δ.

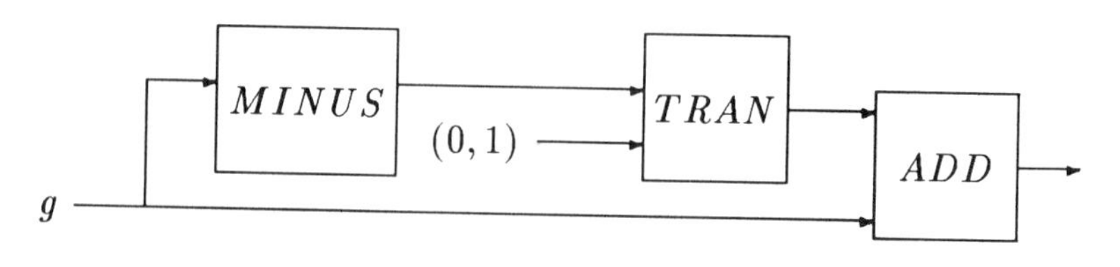

6. Find the first quadrant digital signal g such that

$$g - 4g_{1,1} = \delta_{1,1}$$

7. Check the solution found in Exercise 6. Can convolution be employed for this purpose?

8. Solve for the third quadrant solution g to the difference equation

$$g - 5g_{1,0} = \delta$$

9. Find g as in Exercise 8 where

$$g - 4g_{0,1} = \delta$$

10. Find the first quadrant signal g where

$$g + g_{1,1} = \begin{pmatrix} 2 & 0 \\ 3 & 1 \end{pmatrix}_{0,1}^{0}$$

11. Find the first quadrant signal g where

$$g - g_{1,1} = \begin{pmatrix} 3 & 1 \\ -1 & 0 \end{pmatrix}_{0,1}^{0}$$

12. Use $\mathcal{Z}$ transforms to solve the difference equation in Exercise 10.

13. Use $\mathcal{Z}$ transforms to solve the difference equation in Exercise 11.

14. Use $\mathcal{Z}$ transform techniques to solve for g

$$g - ag_{1,1} = \delta$$

How many solutions can be found using this technique?

6

Wraparound Signal Processing

6.1 Wraparound Signals

Up to now, we dealt primarily with two dimensional digital signals. These signals are defined at all pairs of integers. This is true for signals of finite support as well, since they are zero everywhere except at some finite set of lattice points. We now introduce a new type of signal called the *two dimensional wraparound signal*, or *wraparound signal* for short.

Wraparound signals are real valued and are defined on rectangular subsets of lattice points. Throughout, we will use square regions of lattice points. Specifically, wraparound signals have as their domain the Cartesian product of the first n natural numbers. The domain is the Cartesian product of $\{0, 1, 2, ..., n-1\}$ with itself and is denoted by $\mathbf{Z}_n \times \mathbf{Z}_n$. Accordingly, wraparound signals are signals in $\Re^{Z_n \times Z_n}$. Any signal in $\Re^{Z_n \times Z_n}$ must have a (real number) value at each point (i, j) in

$$\{0, 1, 2, ..., n-1\} \times \{0, 1, 2, ..., n-1\}$$

and undefined elsewhere.

Example 6.1 *Two signals f and g will be given in $\Re^{Z_7 \times Z_7}$. Here, both signals must be defined at each of the given points in the set*

$$\mathbf{Z}_7 \times \mathbf{Z}_7 = \{0, 1, 2, 3, 4, 5, 6\} \times \{0, 1, 2, 3, 4, 5, 6\}.$$

Suppose that

$$f(i, j) = \begin{cases} 2 & (i, j) = (1, 0), (0, 0), (0, 1) \\ 3 & (i, j) = (1, 1) \\ 1 & (i, j) = (2, 2) \\ 0 & otherwise \end{cases}$$

and

$$g(i,j) = \begin{cases} 2 & (i,j) = (1,1), (2,2) \\ -1 & (i,j) = (2,1) \\ 0 & otherwise \end{cases}$$

Note that both f and g are not defined outside of $\mathbf{Z}_7 \times \mathbf{Z}_7$.

Several special signals existing in $\Re^{Z_n \times Z_n}$ will now be given. The constant wraparound signals, which have the same value $a \in \Re$, at each lattice point $(i,j) \in \mathbf{Z}_n \times \mathbf{Z}_n$. These signals are denoted by $a_{Z_n \times Z_n}$ and

$$a_{Z_n \times Z_n}(i,j) = a$$

The wraparound delta, or impulse signal $\delta_{Z_n \times Z_n}$, has value 1 at the origin and is zero valued for the other integers in $\mathbf{Z}_n \times \mathbf{Z}_n$. Thus,

$$\delta(i,j) = \begin{cases} 1 & (i,j) = 0 \\ 0 & (i,j) \in \mathbf{Z}_n \times \mathbf{Z}_n - \{(0,0)\} \end{cases}$$

Since each of these functions are defined on some contiguous rectangular subset of lattice points, it does not make sense to define a unit step type function like the one given in Section 1.2. A unit step in the new environment, $\Re^{Z_n \times Z_n}$, is nothing more than $1_{Z_n \times Z_n}$.

As might be expected, bound matrices can be utilized in representing wraparound signals. Indeed, if $f \in \Re^{Z_n \times Z_n}$ then we will write

$$f = \begin{pmatrix} f(0,n-1) & f(1,n-1) & \cdots & f(n-1,n-1) \\ \cdot & \cdot & & \cdot \\ \cdot & \cdot & & \cdot \\ \cdot & \cdot & & \cdot \\ f(0,0) & f(1,0) & \cdots & f(n-1,0) \end{pmatrix}_{0,n-1}^{wn}$$

or

$$f = \begin{pmatrix} f(0,n-1) & f(1,n-1) & \cdots & f(n-1,n-1) \\ \cdot & \cdot & & \cdot \\ \cdot & \cdot & & \cdot \\ \cdot & \cdot & & \cdot \\ f(0,0) & f(1,0) & \cdots & f(n-1,0) \end{pmatrix}^{wn}$$

The exponent, wn, on the right parenthesis stands for wraparound with n^2 entries. It indicates that the signal is to be understood as a wraparound signal and as such $f(i,j)$ is undefined for all lattice points (i,j) with $i < 0$, $j < 0$, $i \geq n$ or $j \geq n$. Further, more convenient bound matrix representations of wraparound signals will be given subsequently.

Example 6.2 *Refer to the wraparound signals f and g given in Example 6.1. Notice that*

$$f = \begin{pmatrix} 0 & 0 & 0 & 0 & 0 & 0 & 0 \\ 0 & 0 & 0 & 0 & 0 & 0 & 0 \\ 0 & 0 & 0 & 0 & 0 & 0 & 0 \\ 0 & 0 & 0 & 0 & 0 & 0 & 0 \\ 0 & 0 & 1 & 0 & 0 & 0 & 0 \\ 2 & 3 & 0 & 0 & 0 & 0 & 0 \\ 2 & 2 & 0 & 0 & 0 & 0 & 0 \end{pmatrix}_{0,6}^{w7}$$

and

$$g = \begin{pmatrix} 0 & 0 & 0 & 0 & 0 & 0 & 0 \\ 0 & 0 & 0 & 0 & 0 & 0 & 0 \\ 0 & 0 & 0 & 0 & 0 & 0 & 0 \\ 0 & 0 & 0 & 0 & 0 & 0 & 0 \\ 0 & 0 & 2 & 0 & 0 & 0 & 0 \\ 0 & 2 & -1 & 0 & 0 & 0 & 0 \\ 0 & 0 & 0 & 0 & 0 & 0 & 0 \end{pmatrix}_{0,6}^{w7}$$

Example 6.3 *Bound matrix representations will now be given for the special wraparound signals previously discussed. We have*

$$3_{Z_4 \times Z_4} = \begin{pmatrix} 3 & 3 & 3 & 3 \\ 3 & 3 & 3 & 3 \\ 3 & 3 & 3 & 3 \\ 3 & 3 & 3 & 3 \end{pmatrix}_{0,3}^{w4}$$

$$0_{Z_3 \times Z_3} = \begin{pmatrix} 0 & 0 & 0 \\ 0 & 0 & 0 \\ 0 & 0 & 0 \end{pmatrix}_{0,2}^{w3}$$

$$1_{Z_5 \times Z_5} = \begin{pmatrix} 1 & 1 & 1 & 1 & 1 \\ 1 & 1 & 1 & 1 & 1 \\ 1 & 1 & 1 & 1 & 1 \\ 1 & 1 & 1 & 1 & 1 \\ 1 & 1 & 1 & 1 & 1 \end{pmatrix}_{0,4}^{w5}$$

$$\delta_{Z_3 \times Z_3} = \begin{pmatrix} 0 & 0 & 0 \\ 0 & 0 & 0 \\ 1 & 0 & 0 \end{pmatrix}_{0,2}^{w3}$$

Similar to the digital signal situation, we define the co-zero set for f in $\Re^{Z_n \times Z_n}$ to be that part of $\mathbf{Z}_n \times \mathbf{Z}_n$ for which f does not take on the value zero. Accordingly, if we denote this set by $coz(f)$ then

$$coz(f) = f^{-1}(\Re - \{0,0\})$$

Additionally, $supp(f)$ is used in denoting a support set of f, that is $supp(f)$ is any set in $\mathbf{Z}_n \times \mathbf{Z}_n$ which contains $coz(f)$.

Example 6.4 *Again refer to the wraparound signals f and g given in Example 6.1. Here,*

$$coz(f) = \{(1,0),(0,0),(0,1),(1,1),(2,2)\} \subset \mathbf{Z}_7 \times \mathbf{Z}_7$$

and

$$coz(g) = \{(1,1),(2,2),(2,1)\} \subset \mathbf{Z}_7 \times \mathbf{Z}_7$$

The co-zero set will be instrumental when wraparound convolution and correlation are described in the sequential as well as, parallel environments. Additionally, compressed versions of bound matrices will be given for various wraparound signals.

6.2 Range Induced Operations and Terms for Wraparound Signals

Signals in $\Re^{Z_n \times Z_n}$ have the reals as their codomain, consequently the range induced operations existing in $\Re^{Z_n \times Z_n}$ are, in essence, identical to those existing in $\Re^{Z \times Z}$. Accordingly, we will employ the same notation as in Section 2.1 to denote wraparound range induced operations and their terms.

As in $\Re^{Z\times Z}$, there is an addition, multiplication, and maximum operation. These are denoted by: ADD, $MULT$, and MAX, or they are equivalently denoted by $+$, $\cdot$, and $\vee$, respectively. Each operator takes two wraparound signals and returns a third wraparound signal. The operations are defined pointwise:

$$ADD(f,g)(i,j) = (f+g)(i,j) = f(i,j) + g(i,j) \qquad (i,j) \in \mathbf{Z}_n \times \mathbf{Z}_n$$

$$MULT(f,g)(i,j) = (f \cdot g)(i,j) = f(i,j) \cdot g(i,j) \qquad (i,j) \in \mathbf{Z}_n \times \mathbf{Z}_n$$

$$MAX(f,g)(n) = (f \vee g)(i,j) = f(i,j) \vee g(i,j) \qquad (i,j) \in \mathbf{Z}_n \times \mathbf{Z}_n$$

As before,

$$coz(f \cdot g) = coz(f) \cap coz(g)$$

and

$$\text{supp}(f+g) \subset coz(f) \cup coz(g)$$

Block diagrams illustrating these operations are identical to those previously given in the non-wraparound situation.

Example 6.5 *Consider f and g given below.*

$$f = \begin{pmatrix} 0 & 3 & 2 & 0 \\ 0 & 2 & 4 & 0 \\ 0 & 0 & 0 & 0 \\ 0 & 3 & -1 & 0 \end{pmatrix}_{0,3}^{w4}$$

$$g = \begin{pmatrix} 0 & 0 & 0 & 0 \\ 0 & -1 & 2 & 3 \\ 0 & 0 & 1 & 2 \\ 0 & 2 & -1 & 0 \end{pmatrix}_{0,3}^{w4}$$

We have, using block diagrams:

f, g → ADD → $f+g =$

$$\begin{pmatrix} 0 & 3 & 2 & 0 \\ 0 & 1 & 6 & 3 \\ 0 & 0 & 1 & 2 \\ 0 & 5 & -2 & 0 \end{pmatrix}_{0,3}^{w4}$$

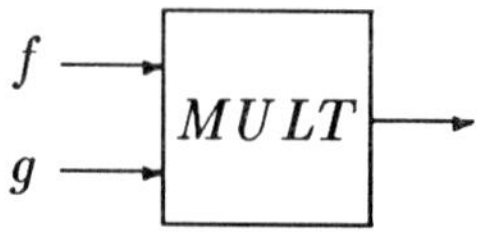

$$f, g \rightarrow \boxed{MULT} \rightarrow f \cdot g = \begin{pmatrix} 0 & 0 & 0 & 0 \\ 0 & -2 & 8 & 0 \\ 0 & 0 & 0 & 0 \\ 0 & 6 & 1 & 0 \end{pmatrix}_{0,3}^{w4}$$

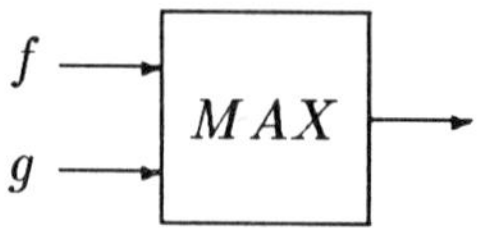

$$f, g \rightarrow \boxed{MAX} \rightarrow f \vee g = \begin{pmatrix} 0 & 3 & 2 & 0 \\ 0 & 2 & 4 & 3 \\ 0 & 0 & 1 & 2 \\ 0 & 3 & -1 & 0 \end{pmatrix}_{0,3}^{w4}$$

Additionally, notice that

$$coz(f \cdot g) = \{(1,0), (2,0), (1,2), (2,2)\} = coz(f) \cap coz(g)$$

Wraparound signals can be thought to exist on a donut, or genus one type surface. Consequently, more compressed representations often can be given for these types of signals. For instance, in the last example notice that f can be written as

$$f = \begin{pmatrix} 3 & 2 \\ 2 & 4 \\ 0 & 0 \\ 3 & -1 \end{pmatrix}_{1,3}^{w4}$$

which can also be written as

$$f = \begin{pmatrix} 3 & -1 \\ 3 & 2 \\ 2 & 4 \end{pmatrix}_{1,0}^{w4}$$

All the range induced operations defined in Sections 2.1 and 2.2 have analogs in $\Re^{Z_n \times Z_n}$. Again, the same notation shall be employed. To begin, the square law "device", denoted by SQ or $()^2$ and the negation operation, denoted by $MINUS$ or by "$-$", are unary operations defined pointwise by

$$SQ(f)(i,j) = f^2(i,j) \qquad (i,j) \in \mathbf{Z}_n \times \mathbf{Z}_n$$

$$MINUS(f)(i,j) = -f(i,j) \qquad (i,j) \in \mathbf{Z}_n \times \mathbf{Z}_n$$

The minimum operation is a binary operation denoted by MIN or $\wedge$, and is defined by

$$MIN(f,g)(i.j) = f(i,j) \wedge g(i,j) \qquad (i,j) \in \mathbf{Z}_n \times \mathbf{Z}_n$$

There are three rectifier type unary devices: the absolute value, the positive part, and the negative part operation, denoted by ABS, POS and NEG, respectively. Alternate notation for the ABS, POS and NEG operations are $| f |$, $(f)^+$ and $(f)^-$, respectively. They are defined by

$$| f | (i,j) = | f(i,j) | \qquad (i,j) \in \mathbf{Z}_n \times \mathbf{Z}_n$$

$$(f)^+ (i,j) = \begin{cases} f(i,j) & f(i,j) > 0 \quad (i,j) \in \mathbf{Z}_n \times \mathbf{Z}_n \\ 0 & f(i,j) \leq 0 \quad (i,j) \in \mathbf{Z}_n \times \mathbf{Z}_n \end{cases}$$

$$(f)^- (i,j) = \begin{cases} -f(i,j) & f(i,j) < 0 \quad (i,j) \in \mathbf{Z}_n \times \mathbf{Z}_n \\ 0 & f(i,j) \geq 0 \quad (i,j) \in \mathbf{Z}_n \times \mathbf{Z}_n \end{cases}$$

Scalar multiplication and offset operations also exist in $\Re^{Z_n \times Z_n}$ and are denoted by SCALAR and OFFSET. When applied to a wraparound signal f and a real number a, the alternate notation af and $f + a$ are often employed. These operations are defined by

$$\begin{array}{ll} SCALAR(f;a)(i,j) = af(i,j) & (i,j) \in \mathbf{Z}_n \times \mathbf{Z}_n \\ OFFSET(f;a)(i,j) = f(i,j) + a & (i,j) \in \mathbf{Z}_n \times \mathbf{Z}_n \end{array}$$

All of the operations above are terms. The proof of these facts are identical to the digital signal situation. Moreover, block diagrams illustrating these operations are the same as the ones given earlier in Sections 2.2 and 2.3.

Example 6.6 *Again let*

$$f = \begin{pmatrix} 0 & 3 & 2 & 0 \\ 0 & 2 & 4 & 0 \\ 0 & 0 & 0 & 0 \\ 0 & 3 & -1 & 0 \end{pmatrix}_{0,3}^{w4}$$

and

$$g = \begin{pmatrix} 0 & 0 & 0 & 0 \\ 0 & -1 & 2 & 3 \\ 0 & 0 & 1 & 2 \\ 0 & 2 & -1 & 0 \end{pmatrix}_{0,3}^{w4}$$

Then

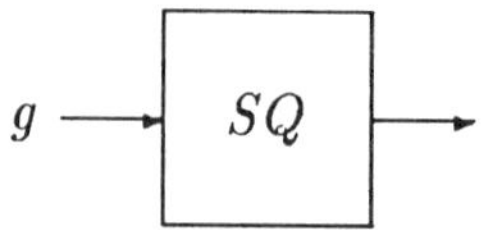

$$g \longrightarrow \boxed{SQ} \longrightarrow (g)^2 = \begin{pmatrix} 0 & 0 & 0 & 0 \\ 0 & 1 & 4 & 9 \\ 0 & 0 & 1 & 4 \\ 0 & 4 & 1 & 0 \end{pmatrix}_{0,3}^{w4}$$

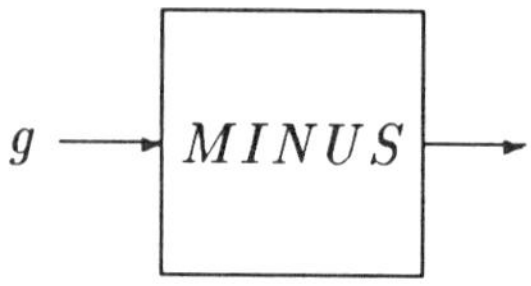

$$g \longrightarrow \boxed{MINUS} \longrightarrow -g = \begin{pmatrix} 0 & 0 & 0 & 0 \\ 0 & 1 & -2 & -3 \\ 0 & 0 & -1 & -2 \\ 0 & -2 & 1 & 0 \end{pmatrix}_{0,3}^{w4}$$

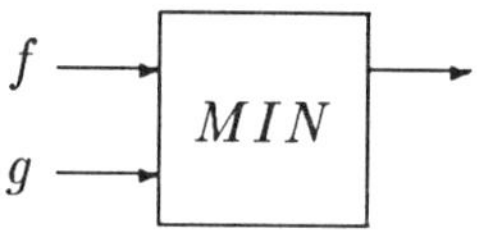

$$f, g \longrightarrow \boxed{MIN} \longrightarrow f \wedge g = \begin{pmatrix} 0 & 0 & 0 & 0 \\ 0 & -1 & 2 & 0 \\ 0 & 0 & 0 & 0 \\ 0 & 2 & -1 & 0 \end{pmatrix}_{0,3}^{w4}$$

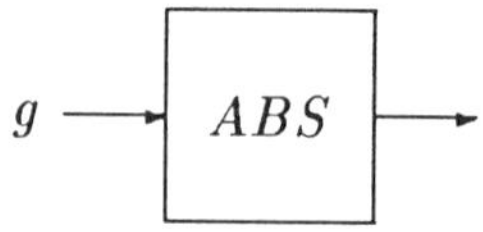

$$g \longrightarrow \boxed{ABS} \longrightarrow |\, g \,| = \begin{pmatrix} 0 & 0 & 0 & 0 \\ 0 & 1 & 2 & 3 \\ 0 & 0 & 1 & 2 \\ 0 & 2 & 1 & 0 \end{pmatrix}_{0,3}^{w4}$$

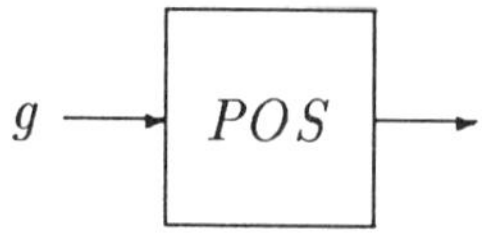

$$g \longrightarrow \boxed{POS} \longrightarrow (g)^+ = \begin{pmatrix} 0 & 0 & 0 & 0 \\ 0 & 0 & 2 & 3 \\ 0 & 0 & 1 & 2 \\ 0 & 2 & 0 & 0 \end{pmatrix}_{0,3}^{w4}$$

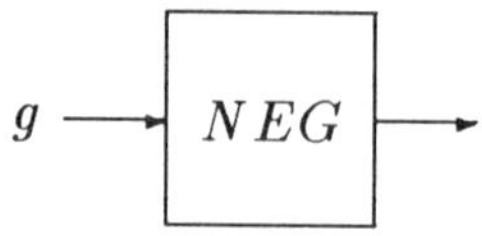

$$g \longrightarrow \boxed{NEG} \longrightarrow (g)^- = \begin{pmatrix} 0 & 0 & 0 & 0 \\ 0 & 1 & 0 & 0 \\ 0 & 0 & 0 & 0 \\ 0 & 0 & 1 & 0 \end{pmatrix}_{0,3}^{w4}$$

$$g, 2 \rightarrow \boxed{SCALAR} \rightarrow 2g = \begin{pmatrix} 0 & 0 & 0 & 0 \\ 0 & -2 & 4 & 6 \\ 0 & 0 & 2 & 4 \\ 0 & 4 & -2 & 0 \end{pmatrix}_{0,3}^{w4}$$

$$g, 2 \rightarrow \boxed{OFFSET} \rightarrow g + 2 = \begin{pmatrix} 0 & 0 & 0 & 0 \\ 0 & 1 & 4 & 5 \\ 0 & 0 & 3 & 4 \\ 0 & 4 & 1 & 0 \end{pmatrix}_{0,3}^{w4}$$

All the range induced algebraic properties holding in $\Re^{Z\times Z}$ also hold in $\Re^{Z_n\times Z_n}$. Accordingly, $\Re^{Z_n\times Z_n}$ is a distributive lattice, commutative, associative algebra with identity (see Section 2.4).

6.3 Domain Induced Operations for Wraparound Signals

Wraparound signals have as their domain the set $\mathbf{Z}_n\times\mathbf{Z}_n$. Similar to the set of lattice points $\mathbf{Z} \times \mathbf{Z}$, group operations can be defined in $\mathbf{Z}_n\times\mathbf{Z}_n$. The group operations in the $\mathbf{Z}_n\times\mathbf{Z}_n$ domain induce similar operations in $\Re^{Z_n\times Z_n}$, which result in the wraparound phenomena. In order to fully understand these domain induced operations, we must define a form of addition and a negation operation in $\mathbf{Z}_n$. Before this is done, however, it is useful to introduce the successor function in $\mathbf{Z}_n$, which is really addition with one argument equal to one. Moreover, it induces the shift operation on signals which clearly illustrates the cyclic or wraparound methodology.

In $\mathbf{Z}_n$ we will define a wraparound type of successor function s. The successor of any number in $\mathbf{Z}_n$, except for $n-1$ is the usual successor - the number which is greater by one. For $n-1$ the successor is zero. So

$$s(k) = \begin{cases} k+1 & k \neq n-1 \\ 0 & k = n-1 \end{cases}$$

One could imagine the numbers $0, 1, ..., n-1$ lying on a circle, equidistant apart; the successor of any integer is found using the integer one unit clockwise.

It is useful to view the successor operation as a type of addition. Indeed, $s(k)$ is the least nonnegative remainder obtained by dividing n into the sum of k plus 1. The reason for this interpretation of the successor function will become evident below. It relates the successor operation with a simple instance of a more general addition type operation in $\mathbf{Z}_n$.

Addition involving integers h and k in $\mathbf{Z}_n$ is denoted by h [+] k and is often called addition modulo n. It is found by taking the remainder of the division by n of the usual sum of h and k. More precisely,

$$h\ [+]\ k = r\ , \text{ where } 0 \leq r < n$$

and r is found by setting

$$r = h + k - nq, \quad \text{using } q = 0 \text{ or } 1, \text{ whichever "works"}$$

i.e. whichever gives r in the desired range $\{0, 1, 2, 3, ..., n-1\}$.

Example 6.7 *Suppose that we are given*

$$\mathbf{Z}_7 = \{0, 1, 2, 3, 4, 5, 6\}$$

then

$$3\ [+]\ 2 = 5$$

because

$$5 = 3 + 2 - 7 \cdot 0$$

and

$$4\ [+]\ 6 = 3$$

note that

$$4 + 6 - 7 \cdot 0 = 10$$

but

$$4 + 6 - 7 \cdot 1 = 3$$

Additionally,

$$5\ [+]\ 6 = 4$$

Notice that the sum h [+] k can be obtained using the successor function repeatedly. Indeed,

$$h\,[+]\,k = s(s(...(s(h)))) = s^k(h)$$

also

$$h\,[+]\,k = s(s(...(s(k)))) = s^h(k)$$

As a consequence, the addition operation satisfies the associative laws. Furthermore, we see

$$0\,[+]\,k = k$$

Example 6.8 *In* $\mathbf{Z}_9$

$$7\,[+]\,3 = 1$$

additionally, we also have

$$s^3(7) = s(s(s(7))) = s(s(8)) = s(0) = 1.$$

Another operation in $\mathbf{Z}_n$ is the minus function $[-]$. For every k in $\mathbf{Z}_n$

$$[-]k = \begin{cases} 0 & \text{if } k = 0 \\ n - k & \text{otherwise} \end{cases}$$

The minus operation can also be found similar to the addition function previously given. Indeed,

$$[-]\,k = r$$

where r is found by setting

$$r = 0 - k - nq$$

and using $q = 0$ or -1, whichever gives a value for r in $\mathbf{Z}_n$. As a consequence,

$$([-]\,k)\,[+]\,k = 0$$

We also write $h\,[-]\,k$ instead of $h\,[+]\,[-]\,k$.

Example 6.9 *In* $\mathbf{Z}_9$ *notice that*

$$[-]3 = 6, \text{ and that } [-]8 = 1$$

whereas

$$[-]1 = 8$$

The set $\mathbf{Z}_n$ with the aforementioned operations and properties form a cyclic group. This structure induces operations in $\mathbf{Z}_n \times \mathbf{Z}_n$, which then domain induces operations in $\Re^{Z_n \times Z_n}$. The first operation to be considered herein is the shift operation denoted by S and

$$S : \Re^{Z_n \times Z_n} \longrightarrow \Re^{Z_n \times Z_n}$$

It is defined using the successor operation s. Indeed,

$$(S(f))(s(i), j) = f(i, j) \quad \text{where } (i, j) \in \mathbf{Z}_n \times \mathbf{Z}_n$$

In particular,

$$(S(f))(1, k) = f(0, k)$$

That is, the height of $S(f)$ at the point $(1, k)$ equals the height of f at the point $(0, k)$. Also,

$$\begin{gathered}
(S(f))(2, k) = f(1, k) \\
(S(f))(3, k) = f(2, k) \\
\cdot \\
\cdot \\
\cdot \\
(S(f))(n-1, k) = f(n-2, k) \\
(S(f))(0, k) = f(n-1, k)
\end{gathered}$$

Thus, we see that this operation translates the signal f to the right. That is, it moves the heights of f point by point, one unit to the right except for the values in the last column, that is, points $(n-1, k)$. These values get wrapped around to become the values of $S(f)$ in the first column, that is the points $(0, k)$.

The operator S can be applied in succession. Accordingly, we write

$$S^2(f) = S(S(f))$$

and

$$S^3(f) = S(S(S(f)))$$

and so on:

$$S^k(f) = S\,S...(S(f))$$

Clearly,

$$S^n(f) = f$$

Example 6.10 *Consider the signal*

$$f = \begin{pmatrix} 1 & 0 & 2 & 3 \\ 0 & 0 & 0 & 1 \\ 0 & 1 & 0 & 0 \\ 4 & 0 & 0 & -1 \end{pmatrix}_{0,3}^{w4}$$

in $\Re^{Z_4 \times Z_4}$. *The shift operation applied to* f *gives*

$$S(f) = \begin{pmatrix} 3 & 1 & 0 & 2 \\ 1 & 0 & 0 & 0 \\ 0 & 0 & 1 & 0 \\ -1 & 4 & 0 & 0 \end{pmatrix}_{0,3}^{w4}$$

Applying the shift again yields

$$S^2(f) = \begin{pmatrix} 2 & 3 & 1 & 0 \\ 0 & 1 & 0 & 0 \\ 0 & 0 & 0 & 1 \\ 0 & -1 & 4 & 0 \end{pmatrix}_{0,3}^{w4}$$

and

$$S^3(f) = \begin{pmatrix} 0 & 2 & 3 & 1 \\ 0 & 0 & 1 & 0 \\ 1 & 0 & 0 & 0 \\ 0 & 0 & -1 & 4 \end{pmatrix}_{0,3}^{w4}$$

Moreover,

$$S^4(f) = f$$

The block diagram illustrating the shift operation is

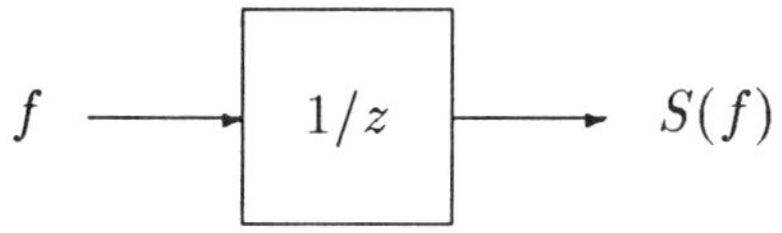

$f \longrightarrow$ [$1/z$] $\longrightarrow S(f)$

As in the case of digital signals, there exists a $\mathcal{Z}$ type transform for wraparound signals. This is discussed in Section 6.7. The symbol $1/z$ illustrates a shift to the right.

Example 6.11 *Suppose that*

$$f = \begin{pmatrix} 3 & 2 \\ 4 & 1 \end{pmatrix}_{4,2}^{w6}$$

is given in $\Re^{Z_6 \times Z_6}$. *Then*

$$f \longrightarrow \boxed{\;1/z\;} \longrightarrow S(f) = \begin{pmatrix} 2 & 0 & 0 & 0 & 0 & 3 \\ 1 & 0 & 0 & 0 & 0 & 4 \end{pmatrix}_{0,2}^{w6} = \begin{pmatrix} 3 & 2 \\ 4 & 1 \end{pmatrix}_{5,2}^{w6}$$

Notice that the last bound matrix for $S(f)$ *is minimal.*

The next domain induced operation is the 90° rotation, denoted by N or by $NINETY$ where

$$N : \Re^{Z_n \times Z_n} \longrightarrow \Re^{Z_n \times Z_n}$$

It is defined by

$$N(f)(i, j) = f(j, [-]\, i)$$

Example 6.12 *Consider the wraparound signal*

$$f = \begin{pmatrix} 3 & 2 \\ 4 & 1 \end{pmatrix}_{4,2}^{w6}$$

given in Example 6.11. *In* $\mathbf{Z}_6$ *we have*

$$\begin{aligned} N(f)(4,4) &= f(4,2) = 3 \\ N(f)(5,4) &= f(4,1) = 4 \\ N(f)(4,5) &= f(5,2) - 2 \\ N(f)(5,5) &= f(5,1) = 1 \end{aligned}$$

Thus,

$$N(f) = \begin{pmatrix} 2 & 1 \\ 3 & 4 \end{pmatrix}_{4,5}^{w6}$$

Notice that if $N(f)$ is shifted one time we obtain

$$S(N(f)) = \begin{pmatrix} 1 & 2 \\ 4 & 3 \end{pmatrix}_{5,5}^{w6}$$

The following block diagram illustrates the operation N on the wraparound signal f.

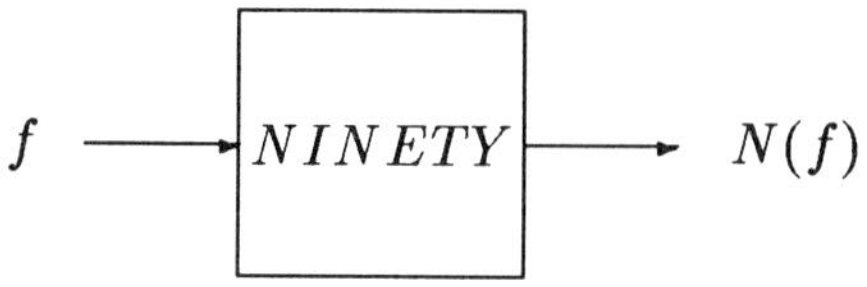

$$f \longrightarrow \boxed{NINETY} \longrightarrow N(f)$$

Example 6.13 *Let*

$$f = \begin{pmatrix} 1 & 0 & 2 & 3 \\ 0 & 0 & 0 & 1 \\ 0 & 1 & 0 & 0 \\ 4 & 0 & 0 & -1 \end{pmatrix}_{0,3}^{w4}$$

as in Example 6.10. *Then we have*

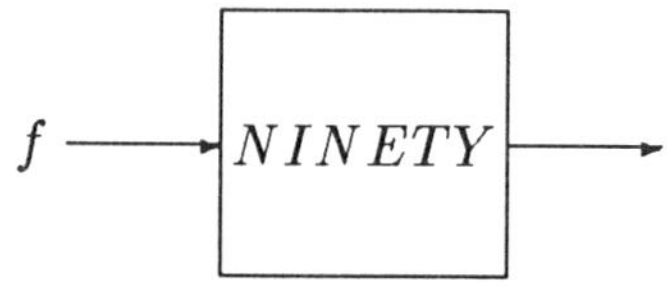

$$f \longrightarrow \boxed{NINETY} \longrightarrow N(f) = \begin{pmatrix} -1 & 3 & 1 & 0 \\ 0 & 2 & 0 & 0 \\ 0 & 0 & 0 & 1 \\ 4 & 1 & 0 & 0 \end{pmatrix}_{0,3}^{w4}$$

The final fundamental domain induced operation to be introduced herein is the diagonal reflection operation, denoted by $REFLECT$ or by D. It too is a unary operation:

$$D : \Re^{Z \times Z} \longrightarrow \Re^{Z \times Z}$$

where

$$D(f)(i,j) = f([-]j, [-]i)$$

As in the digital signal case, this operation is similar to a matrix transpose operation, however, due to the wraparound phenomena, certain rows or columns will be shifted (wrapped around). This operation is illustrated by the block diagram

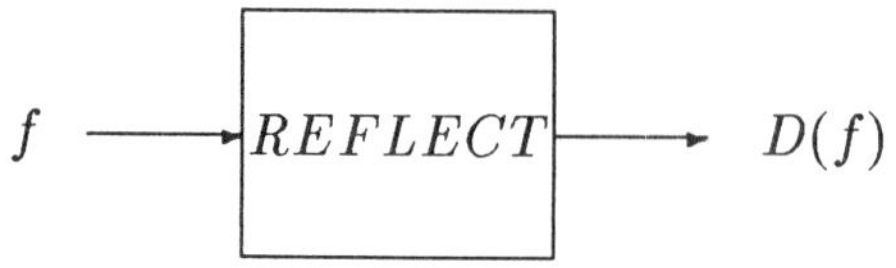

$$f \longrightarrow \boxed{REFLECT} \longrightarrow D(f)$$

Example 6.14 *Again, let*

$$f = \begin{pmatrix} 3 & 2 \\ 4 & 1 \end{pmatrix}^{w6}_{4,2}$$

then

$$\begin{aligned} D(f)(4,2) &= f(4,2) = 3 \\ D(f)(5,2) &= f(4,1) = 4 \\ D(f)(4,1) &= f(5,2) = 2 \\ D(f)(5,1) &= f(5,1) = 1 \end{aligned}$$

and so

$$D(f) = \begin{pmatrix} 3 & 4 \\ 2 & 1 \end{pmatrix}^{w6}_{4,2}$$

Example 6.15 *As in Example* 6.13, *let*

$$f = \begin{pmatrix} 1 & 0 & 2 & 3 \\ 0 & 0 & 0 & 1 \\ 0 & 1 & 0 & 0 \\ 4 & 0 & 0 & -1 \end{pmatrix}^{w4}_{0,3}$$

then

$$D(f) = \begin{pmatrix} 0 & 0 & 0 & 1 \\ 0 & 2 & 0 & 0 \\ -1 & 3 & 1 & 0 \\ 4 & 1 & 0 & 0 \end{pmatrix}^{w4}_{0,3}$$

The macro wraparound operation of translation is denoted by $TRANS$ or by T. It is a shifting type operation which can move the signal f to the right, left, up, or down by an arbitrary number of units. Thus,

$$T : \Re^{Z\times Z} \times \mathbf{Z} \times \mathbf{Z} \longrightarrow \Re^{Z\times Z}$$

where

$$T(f;p,q)(i,j) = f(i\,[-]\,p, j\,[-]\,q)$$

The fact that $T(f;p,q)$ is a term follows from the expression

$$T(f;p,q) = \begin{cases} NS^qN^3S^p(f) & p \geq 0,\ q \geq 0 \\ N^3S^{-q}NS^p(f) & p \geq 0,\ q < 0 \\ NS^qNS^{-p}N^2(f) & p < 0,\ q \geq 0 \\ N^3S^{-q}N^3S^{-p}N^2(f) & p < 0,\ q \leq 0 \end{cases}$$

The quantities $-p$, $-q$, p, and q should be evaluated modulo n. We also denote $T(f;p,q)$ by $f_{p,q}$.

Example 6.16 *Suppose that the wraparound signal f in $\Re^{Z_7 \times Z_7}$ is given by*

$$f = \begin{pmatrix} 2 & 0 & 0 & 0 & 0 & 4 & 0 \\ -1 & 0 & 0 & 0 & 0 & 2 & 0 \end{pmatrix}_{0,1}^{w7} = \begin{pmatrix} 4 & 0 & 2 \\ 2 & 0 & -1 \end{pmatrix}_{5,1}^{w7}$$

then

$$f_{1,0} = \begin{pmatrix} 0 & 2 & 0 & 0 & 0 & 0 & 4 \\ 0 & -1 & 0 & 0 & 0 & 0 & 2 \end{pmatrix}_{0,1}^{w7} = \begin{pmatrix} 4 & 0 & 2 \\ 2 & 0 & -1 \end{pmatrix}_{6,1}^{w7}$$

and

$$f_{-1,0} = S^6(f) = \begin{pmatrix} 0 & 0 & 0 & 0 & 4 & 0 & 2 \\ 0 & 0 & 0 & 0 & 2 & 0 & -1 \end{pmatrix}_{0,1}^{w7} = \begin{pmatrix} 4 & 0 & 2 \\ 2 & 0 & -1 \end{pmatrix}_{4,1}^{w7}$$

Notice that

$$f_{0,1} = \begin{pmatrix} 4 & 0 & 2 \\ 2 & 0 & -1 \end{pmatrix}_{5,2}^{w7}$$

and

$$f_{0,-1} = \begin{pmatrix} 4 & 0 & 2 \\ 2 & 0 & -1 \end{pmatrix}_{5,0}^{w7}$$

The translation operation is denoted using a block diagram by

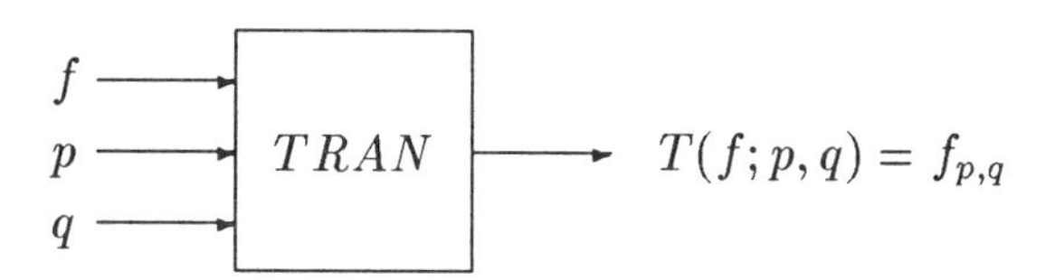

or by

f, (p,q) → TRAN → $f_{p,q}$

6.4 Set Morphology in $\mathbf{Z_n} \times \mathbf{Z_m}$

Several set and arithmetic operations will be described in $\mathbf{Z}_n \times \mathbf{Z}_m$. These operations are needed for a full understanding of more advanced wraparound signal processing operations. Some of which are presented in the next several sections. These operations employ wraparound set morphology in some form.

Four fundamental operations involving subsets of $\mathbf{Z}_n \times \mathbf{Z}_m$ provide the setting to all binary wraparound morphology. The four operations are: set theoretic union, set theoretic intersection, translation, and 180° rotation. If A and B are subsets of $\mathbf{Z}_n \times \mathbf{Z}_m$, then the union and intersection are defined respectively by

$$A \cup B = \{(p,q) \mid (p,q) \in A \text{ or } (p,q) \in B\}$$

$$A \cap B = \{(p,q) \mid (p,q) \in A \text{ and } (p,q) \in B\}$$

The translation operation involving the set A and point $(r,s) \in \mathbf{Z}_n \times \mathbf{Z}_m$ is denoted by $A+(r,s)$. Equivalent notations are $TRAN(A;r,s)$ and $T(A;r,s)$. It is defined by

$$A + (r,s) = \{(r+p, s+q) \quad \text{where } (p,q) \in A\}$$

The integer $r+p$ must lie in $\mathbf{Z}_n$. Accordingly, the addition of $r+p$ and $s+q$ are performed modulo n and modulo m, respectively. Throughout this section, no special notation is used for modulo addition and subtraction.

The sets A and $A+(r,s)$ are identical in structure with the exception that the latter set is A moved r units to the right and s units up. Due to the cyclic properties of $\mathbf{Z}_n$ and $\mathbf{Z}_m$ the translation operation occurs using wraparound, thus a set and its translates may not appear to be identical. This phenomena is illustrated in the example below, and is similar to what we have seen earlier for signals.

The 180° rotation of the set A is denoted by $-A$ and is defined by

$$-A = \{(-r,-s) \quad \text{where } (r,s) \in A\}$$

The negation operations $-r$ and $-s$ are constructed using modulo n and modulo m arithmetic respectively. As in the translation operation, wraparound occurs and the appearance of $-A$ may not be what is intuitively expected.

Example 6.17 *Suppose we are given the set*

$$A = \{(0,2),(1,2),(1,3),(2,2),(2,3),(2,4),(3,4),(3,5)\}$$

in $\mathbf{Z}_6 \times \mathbf{Z}_8$. *The translate*

$$A + (2,1) = \{(2,3),(3,3),(4,3),(3,4),(4,4),(4,5),(5,5),(5,6)\}$$

The translate

$$A + (4,0) = \{(4,2),(5,2),(5,3),(6,2),(6,3),(6,4),(7,4),(7,5)\}$$

However, since 6 *in* $\mathbf{Z}_6$ *is* 0 *and* 7 *is* 1, *we obtain*

$$A + (4,0) = \{(4,2),(5,2),(5,3),(0,2),(0,3),(0,4),(1,4),(1,5)\}$$

Example 6.18 *Again consider the set*

$$A = \{(0,2),(1,2),(1,3),(2,2),(2,3),(2,4),(3,4),(3,5)\}$$

in $\mathbf{Z}_6 \times \mathbf{Z}_8$. *This time we will find* $-A$. *So negating each element in* A *using the modulo minus operation gives*

$$-(0,2) = (0,6) \quad \textit{since} \ -0 = 0$$

and

$$-2 = 8 - 2 = 6 \quad \textit{in } \mathbf{Z}_8$$

$$-(1,2) = (5,6) \quad \textit{since} \ -1 = 6 - 1 = 5 \ \textit{in } \mathbf{Z}_6$$

and so on, then

$$-A = \{(0,7),(5,6),(5,5),(4,6),(4,5),(4,4),(3,4),(3,3)\}$$

All morphological operations in $\mathbf{Z}_n \times \mathbf{Z}_m$ are formed under function composition using union, intersection, translation, and 180° rotation. To begin, the *dilation* of two subsets A and B in $\mathbf{Z}_n \times \mathbf{Z}_m$ is denoted $\mathcal{D}(A, B)$ or by $A \oplus B$. It is also called Minkowski addition and is defined as the union of all translates of A using points from B, that is

$$\mathcal{D}(A,B) = \bigcup_{(p,q) \in B} (A + (p,q))$$

Note that the translation operation must be performed using modulo arithmetic and furthermore

$$A + (p,q) = \mathcal{D}(A, \{(p,q)\})$$

These concepts are illustrated in the following examples.

Example 6.19 *In* $\mathbf{Z}_6 \times \mathbf{Z}_8$ *let*

$$A = \{(0,0),(1,1)\}$$

and

$$B = \{(5,7)\}$$

then

$$\mathcal{D}(A,B) = \{(5,7),(6,8)\} = \{(0,0),(5,7)\}$$

Example 6.20 *Again in* $\mathbf{Z}_6 \times \mathbf{Z}_8$*, suppose that* $\mathcal{D}(A,B)$ *is desired where*

$$A = \{(0,2),(1,2),(1,3),(2,2),(2,3),(2,4),(3,4),(3,5)\}$$

and

$$B = \{(0,0),(1,0),(1,1)\}.$$

For each point in B *we find the translate of* A *by that point*

$$A + (0,0) = A$$

next

$$A + (1,0) = \{(1,2),(2,2),(3,2),(2,3),(3,3),(3,4),(4,4),(4,5)\}$$

and then

$$A + (1,1) = \{(1,3),(2,3),(3,3),(2,4),(3,4),(3,5),(4,5),(4,6)\}.$$

Finally we form the union of all these translates to obtain

$$\mathcal{D}(A,B) = \{(0,2),(1,2),(1,3),(2,2),(2,3),(3,2),(2,4),$$

$$(3,3),(3,4),(4,4),(3,5),(4,5),(4,6)\}$$

The next morphological operation using sets A and B in $\mathbf{Z}_n \times \mathbf{Z}_m$ is *erosion*, and is denoted by $\mathcal{E}(A,B)$. It is defined by forming the set of all points in $\mathbf{Z}_n \times \mathbf{Z}_m$ for which translates of B by these points are subsets of A, i.e.

$$\mathcal{E}(A,B) = \{(p,q) \mid B + (p,q) \subset A\}$$

An operation related to erosion is Minkowski subtraction, denoted by $A \ominus B$ and defined as

$$A \ominus B = \bigcap_{(p,q)\in B} A + (p,q)$$

The erosion of A by B, can be found using Minkowski subtraction of A by $-B$. That is,

$$\mathcal{E}(A,B) = A \ominus -B = \bigcap_{(p,q)\in -B} A + (p,q)$$

Here $-B$ is the 180° rotation of B.

Example 6.21 *Use A and B in $\mathbf{Z}_6 \times \mathbf{Z}_8$ given in Example 6.20. Suppose that $\mathcal{E}(A,B)$ is to be found first by employing the definition. Notice that*

$$\begin{aligned}
B + (0,2) &= \{(0,2),(1,2),(1,3)\} \subset A \\
B + (1,2) &= \{(1,2),(2,2),(2,3)\} \subset A \\
B + (1,3) &= \{(1,3),(2,3),(2,4)\} \subset A \\
B + (2,4) &= \{(2,4),(3,4),(3,5)\} \subset A
\end{aligned}$$

No other translates of B, other than the four just given are subsets of A. Forming the set consisting of the translation values gives

$$\mathcal{E}(A,B) = \{(0,2),(1,2),(1,3),(2,4)\}$$

The same result can be found using Minkowski subtraction. Indeed, here we first obtain $-B$

$$-B = \{(0,0),(-1,0),(-1,-1)\}.$$

The value -1 in $\mathbf{Z}_6$ is 5, and -1 in $\mathbf{Z}_8$ is 7, hence

$$-B = \{(0,0),(5,0),(5,7)\}.$$

Next, translate A by elements in $-B$. So

$$A + (5,0) = \{(5,2),(6,2),(7,2),(6,3),(7,3),(7,4),(8,4),(8,5)\}.$$

Using modulo 6 arithmetic gives

$$A + (5,0) = \{(5,2),(0,2),(1,2),(0,3),(1,3),(1,4),(2,4),(2,5)\}.$$

Similarly, using modulo 6 *arithmetic in the first tuple and modulo* 8 *in the second gives*

$$A + (5,7) = \{(5,1),(0,1),(1,1),(0,2),(1,2),(1,3),(2,3),(2,4)\}.$$

Forming the intersection of A *with* $A+(5,0)$ *and* $A+(5,7)$ *gives*

$$\mathcal{E}(A,B) = \{(0,2),(1,2),(1,3),(2,4)\}.$$

The *opening* operation $\mathcal{O}(A,B)$ involving sets A and B from $\mathbf{Z}_n \times \mathbf{Z}_m$ is defined by the union of all translates of B which are subsets of A

$$\mathcal{O}(A,B) = \bigcup_{B+(p,q)\subset A} B+(p,q)$$

The translation operation is found using modulo arithmetic. An equivalent way of obtaining the opening is by first employing the erosion and then the dilation operation, that is,

$$\mathcal{O}(A,B) = \mathcal{D}(\mathcal{E}(A,B),B)$$

Example 6.22 *Consider*

$$A = \{(0,2),(1,2),(1,3),(2,2),(2,3),(2,4),(3,4),(3,5)\}$$

and

$$B = \{(0,0),(1,0),(1,1)\}$$

in $\mathbf{Z}_6 \times \mathbf{Z}_8$. *We find the opening of* A *by* B *employing the definition. Using the result established in Example* 6.21 *we see that*

$$\begin{aligned} B+(0,2) &= \{(0,2),(1,2),(1,3)\} \subset A \\ B+(1,2) &= \{(1,2),(2,2),(2,3)\} \subset A \\ B+(1,3) &= \{(1,3),(2,3),(2,4)\} \subset A \\ B+(2,4) &= \{(2,4),(3,4),(3,5)\} \subset A \end{aligned}$$

Then forming the union of these four translated sets gives:

$$\mathcal{O}(A,B) = A.$$

The same result can be obtained by first eroding A *by* B. *Again see Example 6.21, where it was shown that*

$$\mathcal{E}(A,B) = \{(0,2),(1,2),(1,3),(2,4)\}.$$

Next, we dilate $\mathcal{E}(A,B)$ by B. Thus, we form

$$\mathcal{E}(A,B) + (0,0) = \mathcal{E}(A,B)$$

$$\mathcal{E}(A,B) + (1,0) = \{(1,2),(2,2),(2,3),(3,4)\}$$

$$\mathcal{E}(A,B) + (1,1) = \{(1,3),(2,3),(2,4),(3,5)\}.$$

Finally, we form the union of these sets to obtain

$$\mathcal{O}(A,B) = \mathcal{D}(\mathcal{E}(A,B),B) = A$$

The final morphological operation to be introduced is the closing of A by B, and is denoted by $\mathcal{C}(A,B)$. It is found by forming the sets of all points (p,q) for which all translates of B simultaneously contain (p,q) and have nonempty intersection with A. Thus, $(p,q) \in \mathcal{C}(A,B)$ if and only if

$$(B + (r,s)) \cap A \neq \emptyset$$

for every $B + (r,s)$ containing (p,q). Notice by definition, any point (p,q) in A must also be in $\mathcal{C}(A,B)$ and so

$$A \subset \mathcal{C}(A,B).$$

Fortunately, closing can also be found in a more algorithmic fashion. Indeed, it can be found using the 180° rotation along with dilation and erosion. Specifically,

$$\mathcal{C}(A,B) = \mathcal{E}(\mathcal{D}(A,-B),-B)$$

Example 6.23 *Using A and B in $\mathbf{Z}_6 \times \mathbf{Z}_8$ provided in the previous example, this time $\mathcal{C}(A,B)$ will be found. First observe that B contains three points. Thus, there will be precisely three translates of B to consider in investigating if a point (p,q) belongs to $\mathcal{C}(A,B)$. If any one of the three translates of B simultaneously contain (p,q) but does not contain a point from A then (p,q) is not in $\mathcal{C}(A,B)$. The point $(2,5) \notin \mathcal{C}(A,B)$ since*

$$(2,5) \in (B + (1,5))$$

and

$$(B + (1,5)) \cap A = \emptyset.$$

Similarly, no point (p, q) *in* $\mathbf{Z}_6 \times \mathbf{Z}_8$ *is in* $\mathcal{C}(A, B)$ *except if* $(p, q) \in A$. *Hence,* $\mathcal{C}(A, B) = A$. *This result can also be obtained by using*

$$\mathcal{C}(A, B) = \mathcal{E}(\mathcal{D}(A, -B), -B).$$

Since

$$-B = \{(0,0), (5,0), (5,7)\}$$

we first find $\mathcal{D}(A, -B)$. *Accordingly, we obtain all translates of* A *by elements in* $-B$,

$$A + (0,0) = A$$

$$A + (5,0) = \{(5,2), (0,2), (1,2), (0,3), (1,3), (1,4), (2,4), (2,5)\}$$

$$A + (5,7) = \{(5,1), (0,1), (1,1), (0,2), (1,2), (1,3), (2,3), (2,4)\}$$

Forming the union of these three sets gives

$$\mathcal{D}(A, -B) = \{(0,2), (1,2), (2,2), (1,3), (2,3), (2,4), (3,4), (3,5), (5,2), (5,2), (0,3), (1,4), (2,5), (5,1), (0,1), (1,1)\}.$$

Next we erode the set $\mathcal{D}(A, -B)$ *by* $-B$. *Notice that*

$$\begin{aligned}
-B + (0,2) &\subset \mathcal{D}(A, -B)\\
-B + (1,2) &\subset \mathcal{D}(A, -B)\\
-B + (1,3) &\subset \mathcal{D}(A, -B)\\
-B + (2,2) &\subset \mathcal{D}(A, -B)\\
-B + (2,3) &\subset \mathcal{D}(A, -B)\\
-B + (2,4) &\subset \mathcal{D}(A, -B)\\
-B + (3,4) &\subset \mathcal{D}(A, -B)\\
-B + (3,5) &\subset \mathcal{D}(A, -B)
\end{aligned}$$

Forming the union of the eight translation values used above again gives

$$\mathcal{C}(A, B) = A.$$

Thus far the only rotation considered in $\mathbf{Z}_n \times \mathbf{Z}_m$ was the 180° rotation. Moreover, the 180° rotation is involutory, that is

$$-(-A) = A.$$

Now, we will consider other 180° type rotations which are found by using 90° type rotations in succession.

The positive (counter-clockwise) 90° rotation operation N maps sets A in $\mathbf{Z}_n \times \mathbf{Z}_m$ into other sets in $\mathbf{Z}_n \times \mathbf{Z}_m$ using the rule

$$N(A) = \{(-j, i) \quad \text{for } (i, j) \in A\}$$

Example 6.24 *If*

$$A = \{(0,2),(1,4),(5,3)\} \subset \mathbf{Z}_6 \times \mathbf{Z}_5$$

is given and $N(A)$ *is to be found, then*

$$N(A) = \{(-2,0),(-4,1),(-3,5)\}$$

The two tuple in this set must be written using elements in $\mathbf{Z}_6 \times \mathbf{Z}_5$. *Accordingly, using modulo arithmetic, we obtain*

$$N(A) = \{(4,0),(2,1),(3,0)\}$$

If N is applied to $N(A)$ the set A is said to undergo a 180° positive rotation and is denoted by $N^2(A)$. What is of principal interest is that $N^2(A)$ need not equal $-A$ (see Exercise 14). A third application of N to A gives the 270° positive rotation and is denoted by $N^3(A)$. Four consecutive applications of N to A gives $N^4(A)$, this result could differ from A itself. The last observation accounts for the seemingly strange results involving $N^2(A)$ in relationship to $-A$.

The final operation, which we will mention, is the reflection operation $\mathcal{D}$, which can also be applied to a subset A of $\mathbf{Z}_n \times \mathbf{Z}_m$. The set $\mathcal{D}(A)$, which geometrically speaking results as a mirror image of A, is found by rotating the set A 180° about a 45° line through the origin and then performing wraparound. More precisely,

$$\mathcal{D}(A) = \{(-j, -i) \quad \text{where } (i, j) \in A\}$$

Like the rotation operation N, $\mathcal{D}$ also behaves in a strange manner. For instance, $\mathcal{D}^2(A)$ need not equal A. Moreover, it is not 1-1 or onto. When $\mathcal{D}$ and N are applied to $\mathbf{Z}_n \times \mathbf{Z}_n$, for $A = \{(i,j)\}$, we have $N(A) = \{(n-j, i)\}$ and $N^2(A) = \{(n-i, n-j)\} = -A$. Accordingly, $N^4(A) = A$ and $\mathcal{D}^2(A) = A$. Thus, the environment of $\mathbf{Z}_n \times \mathbf{Z}_n$ will be used throughout this chapter with the exception of the last section.

6.5 Sequential and Parallel Formulation of Wraparound Convolution

Similar to digital convolution, wraparound convolution is defined for wraparound signals in a pointwise fashion. If f and g are in $\Re^{Z_n \times Z_n}$ then the convolution of f and g is denoted by $CONV(f, g)$ or by $f * g$, and

$$(f * g)(k, m) = \sum_{j,i=0}^{n-1} f(i, j) g(k - i, m - j)$$

for k, $m = 0, 1, 2, ..., n-1$. Here $k\,[-]\,i$ should be used instead of $k - i$, and $m\,[-]\,j$ instead of $m - j$ to denote arithmetic operations in $\mathbf{Z}_n$. Whenever the co-zero set for f or g is not empty, a more compact formula for convolution is

$$(f*g)(k, m) = \begin{cases} \displaystyle\sum_{\substack{(i,j) \in coz(f) \\ (k-i, m-j) \in coz(g)}} f(i, j) g(k - i, m - j) & (k, m) \in \mathcal{D}(coz(f), coz(g)) \\ 0 & \text{otherwise} \end{cases}$$

In terms of block diagrams, the convolution of f and g is illustrated by

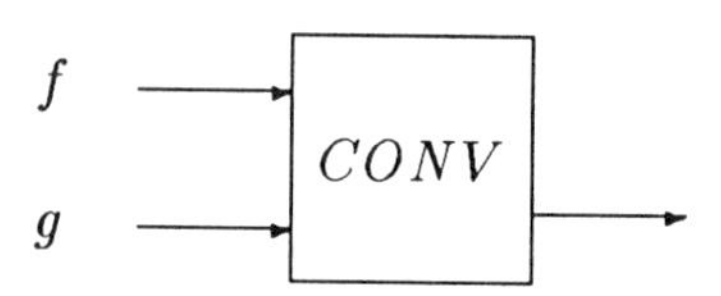

Example 6.25 *Let*

$$f = \begin{pmatrix} 0 & 0 & 0 & 0 \\ 0 & 0 & 0 & 0 \\ 0 & 2 & 1 & 0 \\ 0 & -1 & 3 & 0 \end{pmatrix}_{0,3}^{w4} = \begin{pmatrix} 2 & 1 \\ -1 & 3 \end{pmatrix}_{1,1}^{w4}$$

$$g = \begin{pmatrix} 0 & 0 & 0 & 0 \\ 0 & 0 & 0 & 0 \\ 0 & 0 & 1 & 2 \\ 0 & 0 & 0 & -1 \end{pmatrix}_{0,3}^{w4} = \begin{pmatrix} 1 & 2 \\ 0 & -1 \end{pmatrix}_{2,1}^{w4}$$

The co-zero sets are

$$coz(f) = \{(1,1),(1,0),(2,1),(2,0)\}$$

and

$$coz(g) = \{(2,1),(3,1),(3,0)\}$$

The dilation of $coz(f)$ *and* $coz(g)$ *is given by*

$$\mathcal{D}(coz(f), coz(g)) = \{(3,2),(3,1),(0,2),(0,1),(1,2),(1,1),(0,0),(1,0)\}$$

and so $f * g$ *will be zero outside this set. We will find* $f * g$ *for each point within the dilated set. Beginning with the point* $(3,2)$ *we have*

$$f * g(3,2) = \sum_{\substack{(i,j)\in coz(f)\\ (3,i,2-j)\in coz(g)}} f(i,j)g(3-i,2-j) = 2$$

Similarly, for the other points:

$$f * g(3,1) = f(1,0)g(2,1) = -1$$

$$f * g(0,2) = f(1,1)g(3,1) + f(2,1)g(2,1) = 5$$

$$f * g(0,1) = f(1,1)g(3,0) + f(1,0)g(3,0) + f(2,0)g(2,1) = -1$$

$$f * g(1,2) = f(2,1)g(3,1) = 2$$

$$f * g(1,1) = f(2,1)g(3,0) + f(2,0)g(3,1) = 5$$

$$f * g(0,0) = f(1,0)g(3,0) = 1$$

$$f * g(1,0) = f(2,0)g(3,0) = -3$$

Thus,

$$f * g = \begin{pmatrix} 5 & 2 & 0 & 2 \\ -1 & 5 & 0 & -1 \\ 1 & -3 & 0 & 0 \end{pmatrix}_{0,2}^{w4} = \begin{pmatrix} 2 & 5 & 2 \\ -1 & -1 & 5 \\ 0 & 1 & -3 \end{pmatrix}_{3,2}^{w4}$$

Similar to the case of digital signals as described in Section 3.3, geometric and bound matrix methods can play a greater role in finding wraparound convolutions of f and g. Minimal bound matrix representation shall be employed depending on the size of the co-zero sets of f and g.

If the cardinality of these two sets are small, then we can imitate the method presented in Section 3.3, while taking into account wraparound. If the cardinality of either of these two sets is not small we might want to work with a full $n \times n$ bound matrix. In any case, the first step is to form the 180° rotation of g, thereby obtaining $\tilde{g}$. The next several steps are to form the inner or dot product of f with the translates of $\tilde{g}$. For each translation index of $\tilde{g}$ we form the inner product of f with the translate of $\tilde{g}$. That is, we form the sum of the product of the values of f and $\tilde{g}_{pq}$ pointwise; the result is

$$(f * g)(p, q)$$

Example 6.26 *Again consider*

$$f = \begin{pmatrix} 2 & 1 \\ -1 & 3 \end{pmatrix}_{1,1}^{w4}$$

and

$$g = \begin{pmatrix} 1 & 2 \\ 0 & -1 \end{pmatrix}_{2,1}^{w4}$$

The objective will be to find $f * g$. Notice that

$$\tilde{g} = \begin{pmatrix} 0 & 2 & 1 & 0 \\ 0 & 0 & 0 & 0 \\ 0 & 0 & 0 & 0 \\ 0 & -1 & 0 & 0 \end{pmatrix}_{0,3}^{w4}$$

All possible translates of $\tilde{g}$ will be found such that each translate has some nonzero element in common with

$$f = \begin{pmatrix} 0 & 0 & 0 & 0 \\ 0 & 0 & 0 & 0 \\ 0 & 2 & 1 & 0 \\ 0 & -1 & 3 & 0 \end{pmatrix}_{0,3}^{w4}$$

Only translation indices need be employed contained in the dilated set. At each index of translation of $\tilde{g}$ we multiply the nonzero entries of f and $\tilde{g}$ pointwise and add the result. Using the index of translation $(0,0)$ we obtain

$$(f * g)(0,0) = 1$$

This follows by observing the following bound matrices:

$$\tilde{g}_{0,0} = \begin{pmatrix} 0 & 2 & 1 & 0 \\ 0 & 0 & 0 & 0 \\ 0 & 0 & 0 & 0 \\ 0 & \langle -1 \rangle & 0 & 0 \end{pmatrix}^{w4}_{0,3} \qquad f = \begin{pmatrix} 0 & 0 & 0 & 0 \\ 0 & 0 & 0 & 0 \\ 0 & 2 & 1 & 0 \\ 0 & \langle -1 \rangle & 3 & 0 \end{pmatrix}^{w4}_{0,3}$$

Where $\langle\rangle$ *denotes the position(s) where the value of* f *and the translations of* g *both are not zero, these are the values which determine the inner product.*

$$\tilde{g}_{1,0} = \begin{pmatrix} 0 & 0 & 2 & 1 \\ 0 & 0 & 0 & 0 \\ 0 & 0 & 0 & 0 \\ 0 & 0 & \langle -1 \rangle & 0 \end{pmatrix}^{w4}_{0,3} \qquad f = \begin{pmatrix} 0 & 0 & 0 & 0 \\ 0 & 0 & 0 & 0 \\ 0 & 2 & 1 & 0 \\ 0 & -1 & \langle 3 \rangle & 0 \end{pmatrix}^{w4}_{0,3}$$

gives $(f * g)(1,0) = -3.$

$$\tilde{g}_{0,1} = \begin{pmatrix} 0 & 0 & 0 & 0 \\ 0 & 0 & 0 & 0 \\ 0 & \langle -1 \rangle & 0 & 0 \\ 0 & \langle 2 \rangle & \langle 1 \rangle & 0 \end{pmatrix}^{w4}_{0,3} \qquad f = \begin{pmatrix} 0 & 0 & 0 & 0 \\ 0 & 0 & 0 & 0 \\ 0 & \langle 2 \rangle & 1 & 0 \\ 0 & \langle -1 \rangle & \langle 3 \rangle & 0 \end{pmatrix}^{w4}_{0,3}$$

gives $(f * g)(0,1) = -1.$

$$\tilde{g}_{3,1} = \begin{pmatrix} 0 & 0 & 0 & 0 \\ 0 & 0 & 0 & 0 \\ -1 & 0 & 0 & 0 \\ 2 & \langle 1 \rangle & 0 & 0 \end{pmatrix}^{w4}_{0,3} \qquad f = \begin{pmatrix} 0 & 0 & 0 & 0 \\ 0 & 0 & 0 & 0 \\ 0 & 2 & 1 & 0 \\ 0 & \langle -1 \rangle & 3 & 0 \end{pmatrix}^{w4}_{0,3}$$

gives $(f * g)(3,1) = -1.$

$$\tilde{g}_{3,2} = \begin{pmatrix} 0 & 0 & 0 & 0 \\ -1 & 0 & 0 & 0 \\ 2 & \langle 1 \rangle & 0 & 0 \\ 0 & 0 & 0 & 0 \end{pmatrix}^{w4}_{0,3} \qquad f = \begin{pmatrix} 0 & 0 & 0 & 0 \\ 0 & 0 & 0 & 0 \\ 0 & \langle 2 \rangle & 1 & 0 \\ 0 & -1 & 3 & 0 \end{pmatrix}^{w4}_{0,3}$$

gives $(f * g)(3,2) = 2.$

$$\tilde{g}_{0,2} = \begin{pmatrix} 0 & 0 & 0 & 0 \\ 0 & -1 & 0 & 0 \\ 0 & \langle 2 \rangle & \langle 1 \rangle & 0 \\ 0 & 0 & 0 & 0 \end{pmatrix}^{w4}_{0,3} \qquad f = \begin{pmatrix} 0 & 0 & 0 & 0 \\ 0 & 0 & 0 & 0 \\ 0 & \langle 2 \rangle & \langle 1 \rangle & 0 \\ 0 & -1 & 3 & 0 \end{pmatrix}^{w4}_{0,3}$$

so $(f * g)(0,2) = 5$.

$$\tilde{g}_{1,2} = \begin{pmatrix} 0 & 0 & 0 & 0 \\ 0 & 0 & -1 & 0 \\ 0 & 0 & \langle 2 \rangle & 1 \\ 0 & 0 & 0 & 0 \end{pmatrix}^{w4}_{0,3} \qquad f = \begin{pmatrix} 0 & 0 & 0 & 0 \\ 0 & 0 & 0 & 0 \\ 0 & 2 & \langle 1 \rangle & 0 \\ 0 & -1 & 3 & 0 \end{pmatrix}^{w4}_{0,3}$$

gives $(f * g)(1,2) = 2$.

$$\tilde{g}_{1,1} = \begin{pmatrix} 0 & 0 & 0 & 0 \\ 0 & 0 & 0 & 0 \\ 0 & 0 & \langle -1 \rangle & 0 \\ 0 & 0 & \langle 2 \rangle & 1 \end{pmatrix}^{w4}_{0,3} \qquad f = \begin{pmatrix} 0 & 0 & 0 & 0 \\ 0 & 0 & 0 & 0 \\ 0 & 2 & \langle 1 \rangle & 0 \\ 0 & -1 & \langle 3 \rangle & 0 \end{pmatrix}^{w4}_{0,3}$$

gives $(f * g)(1,1) = 5$. *The result is therefore,*

$$f * g = \begin{pmatrix} 0 & 0 & 0 & 0 \\ 5 & 2 & 0 & 2 \\ -1 & 5 & 0 & -1 \\ 1 & -3 & 0 & 0 \end{pmatrix}^{w4}_{0,3}$$

A parallel algorithm very similar to the one given in Section 3.4 can be employed in finding wraparound convolution. Indeed, the exact same block diagram (Figure 3.1) can be utilized. Of course, when using this diagram in wraparound situations the translation operation must be performed modulo n. All other operations are identical.

Example 6.27 *We will find* $f * g$ *using the parallel convolution operation of Figure* 3.1 *with*

$$f = \begin{pmatrix} 2 & 1 \\ -1 & 3 \end{pmatrix}^{w4}_{1,1}$$

and

$$g = \begin{pmatrix} 1 & 2 \\ 0 & -1 \end{pmatrix}^{w4}_{2,1}$$

Notice that the output of the co-zero block is given as the values in the stack

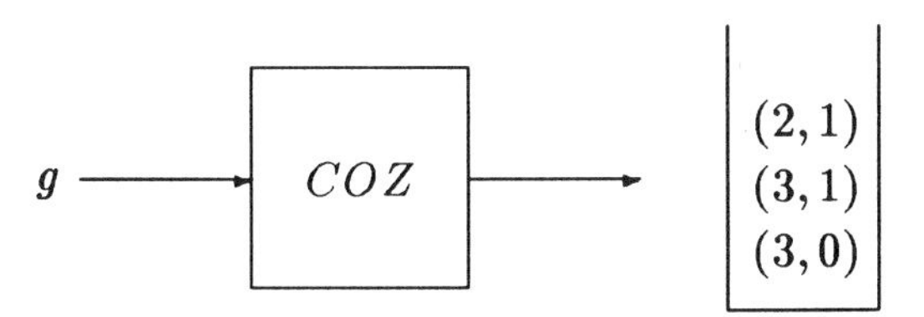

The values in the stack are used as inputs in the transformation blocks. Also, the output of the RANGE block is given by values in the stack

g → RANGE →

1
2
−1

The values in this stack are used as input in the SCALAR blocks. Notice that the three translation scalar block connections provide the following:

$$f_{2,1} = \begin{pmatrix} 2 & 1 \\ -1 & 3 \end{pmatrix}^{w4}_{3,2}$$

$$2f_{3,1} = \begin{pmatrix} 4 & 2 \\ -2 & 6 \end{pmatrix}^{w4}_{4,2} = \begin{pmatrix} 4 & 2 \\ -2 & 6 \end{pmatrix}^{w4}_{0,2}$$

$$-1f_{3,0} = \begin{pmatrix} -2 & -1 \\ 1 & -3 \end{pmatrix}^{w4}_{4,1} = \begin{pmatrix} -2 & -1 \\ 1 & -3 \end{pmatrix}^{w4}_{0,1}$$

These three signals are used as inputs to the add blocks. So,

$$f * g = f_{2,1} + 2f_{3,1} - f_{3,0}$$

$$= \begin{pmatrix} 0 & 0 & 0 & 0 \\ 1 & 0 & 0 & 2 \\ 3 & 0 & 0 & -1 \\ 0 & 0 & 0 & 0 \end{pmatrix}^{w4}_{0,3} + \begin{pmatrix} 0 & 0 & 0 & 0 \\ 4 & 2 & 0 & 0 \\ -2 & 0 & 0 & 0 \\ 0 & 0 & 0 & 0 \end{pmatrix}^{w4}_{0,3} + \begin{pmatrix} 0 & 0 & 0 & 0 \\ 0 & 0 & 0 & 0 \\ -2 & -1 & 0 & 0 \\ 1 & -3 & 0 & 0 \end{pmatrix}^{w4}_{0,3}$$

$$= \begin{pmatrix} 0 & 0 & 0 & 0 \\ 5 & 2 & 0 & 2 \\ -1 & 5 & 0 & -1 \\ 1 & -3 & 0 & 0 \end{pmatrix}^{w4}_{0,3}$$

Banach algebra properties of convolution holding for functions also hold for wraparound signals. Thus, we have for any two dimensional wraparound signals f, g, and h, and any real number a

B1) Associative Law: $f * (g * h) = (f * g) * h$

B2) Distributive Laws:

$$f * (g + h) = f * g + f * h$$
$$(f + g) * h = f * h + g * h$$

B3) Commutative Law: $f * g = g * f$

B4) Identity (Unity) Law: $f * \delta = \delta * f = f$

B5) Scalar Commutativity: $a(f * g) = (af) * g = f * (ag)$

6.6 Sequential and Parallel Formulations of Wraparound Correlation

Wraparound correlation is defined for f and g in $\Re^{Z_n \times Z_n}$ by

$$COR(f,g)(k,m) = \sum_{j,i=0}^{n-1} f(i,j)g(i-k,j-m)$$

for m, $k = 0, 1, 2, ..., n-1$. As in the digital convolution operation, we have the equivalent, possibly more compressed representation

$$COR(f,g)(k,m) = \begin{cases} \displaystyle\sum_{\substack{(i,j)\in coz(f) \\ (i-k,j-m)\in coz(g)}} f(i,j)g(i-k,j-m) & (k,m) \in \mathcal{D}(coz(f), coz(N^2(g))) \\ 0 & \text{otherwise} \end{cases}$$

The following block diagram illustrates the correlation operation

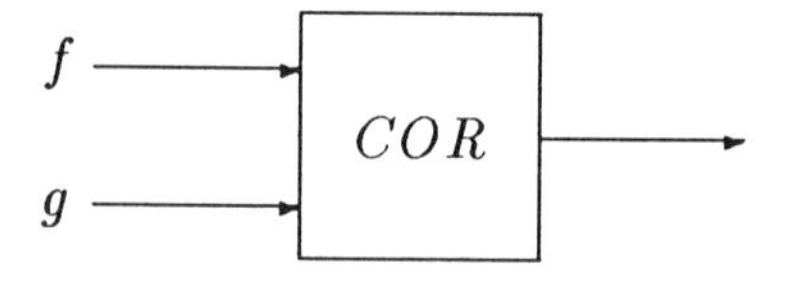

The first step in forming the correlation function $COR(f,g)$ is to find the co-zero sets of f and g, as well as $coz(N^2(g))$. The dilated set $\mathcal{D}(coz(f), coz(N^2(g)))$ must then be formed.

Example 6.28 *Suppose that $COR(f,g)$ is desired, where*

$$f = \begin{pmatrix} 2 & -1 \\ 0 & 1 \end{pmatrix}_{0,1}^{w5}$$

and

$$g = \begin{pmatrix} 0 & 2 \\ 3 & 0 \end{pmatrix}_{0,1}^{w5}$$

Notice that

$$coz(f) = \{(0,1),(1,1),(1,0)\}$$

and

$$coz(g) = \{(0,0),(1,1)\}$$

Then, if $\tilde{g} = N^2(g)$, we obtain,

$$\tilde{g} = \begin{pmatrix} 0 & 3 \\ 2 & 0 \end{pmatrix}_{4,0}^{w5} = \begin{pmatrix} 0 & 0 & 0 & 0 & 2 \\ 0 & 0 & 0 & 0 & 0 \\ 0 & 0 & 0 & 0 & 0 \\ 0 & 0 & 0 & 0 & 0 \\ 3 & 0 & 0 & 0 & 0 \end{pmatrix}_{0,4}^{w5}$$

Thus,

$$coz(\tilde{g}) = \{(0,0),(4,4)\}$$

and so

$$\mathcal{D}(coz(f), coz(\tilde{g})) = \{(0,1),(1,1),(1,0),(4,0),(0,4),(0,0)\}$$

Accordingly, $COR(f,g)$ will be zero everywhere except possibly inside this dilated set. For points in the dilated sets, we have:

$$COR(f,g)(0,1) = \sum_{\substack{(i,j)\in coz(f) \\ (i,j-1)\in coz(g)}} f(i,j)g(i,j-1) = f(0,1)g(0,0) = 6$$

similarly,

$$COR(f,g)(1,1) = f(1,1)g(0,0) = -3$$

$$COR(f,g)(1,0) = f(1,0)g(0,0) = 3$$

$$COR(f,g)(4,0) = f(0,1)g(1,1) = 4$$

$$COR(f,g)(0,4) = f(1,0)g(1,1) = 2$$

$$COR(f,g)(0,0) = f(1,1)g(1,1) = -2$$

Thus,

$$COR(f,g) = \begin{pmatrix} 2 & 0 & 0 & 0 & 0 \\ 0 & 0 & 0 & 0 & 0 \\ 0 & 0 & 0 & 0 & 0 \\ 6 & -3 & 0 & 0 & 0 \\ -2 & 3 & 0 & 0 & 4 \end{pmatrix}^{w5}_{0,4}$$

As in the case of wraparound convolution, wraparound correlation can also be performed with greater geometric insight, and using bound matrices. No 180° rotation need be performed on the signal g when finding $COR(f,g)$ using this method. All that need be done is to find translates of g and form the dot product of these translates with f. The result is the correlation evaluated at the index of translation.

Example 6.29 *As an example, we will find $COR(f,g)$ with*

$$f = \begin{pmatrix} 2 & -1 \\ 0 & 1 \end{pmatrix}^{w5}_{0,1}$$

and

$$g = \begin{pmatrix} 0 & 2 \\ 3 & 0 \end{pmatrix}^{w5}_{0,1}$$

Then, using $g_{0,0}$, forming the dot product of the matrices f and g gives

$$COR(f,g)(0,0) = -2$$

Next, we form the dot product of five other translates of g with f. Thus forming

$$g_{0,1} = \begin{pmatrix} 0 & 0 & 2 & 0 \\ 0 & \langle 3 \rangle & 0 & 0 \\ 0 & 0 & 0 & 0 \\ 0 & 0 & 0 & 0 \end{pmatrix}^{w5}_{4,2} \quad \textit{and} \quad f = \begin{pmatrix} 0 & 0 & 0 & 0 \\ 0 & \langle 2 \rangle & -1 & 0 \\ 0 & 0 & 1 & 0 \\ 0 & 0 & 0 & 0 \end{pmatrix}^{w5}_{4,2}$$

gives

$$COR(f,g)(0,1) = 6.$$

$$g_{1,1} = \begin{pmatrix} 0 & 0 & 0 & 2 \\ 0 & 0 & \langle 3 \rangle & 0 \\ 0 & 0 & 0 & 0 \\ 0 & 0 & 0 & 0 \end{pmatrix}_{4,2}^{w5} \quad and \quad f = \begin{pmatrix} 0 & 0 & 0 & 0 \\ 0 & 2 & \langle -1 \rangle & 0 \\ 0 & 0 & 1 & 0 \\ 0 & 0 & 0 & 0 \end{pmatrix}_{4,2}^{w5}$$

gives $COR(f,g)(1,1) = -3$.

$$g_{1,0} = \begin{pmatrix} 0 & 0 & 0 & 0 \\ 0 & 0 & 0 & 2 \\ 0 & 0 & \langle 3 \rangle & 0 \\ 0 & 0 & 0 & 0 \end{pmatrix}_{4,2}^{w5} \quad and \quad f = \begin{pmatrix} 0 & 0 & 0 & 0 \\ 0 & 2 & -1 & 0 \\ 0 & 0 & \langle 1 \rangle & 0 \\ 0 & 0 & 0 & 0 \end{pmatrix}_{4,2}^{w5}$$

gives $COR(f,g)(1,0) = 3$.

$$g_{0,4} = \begin{pmatrix} 0 & 0 & 0 & 0 \\ 0 & 0 & 0 & 0 \\ 0 & 0 & \langle 2 \rangle & 0 \\ 0 & 3 & 0 & 0 \end{pmatrix}_{4,2}^{w5} \quad and \quad f = \begin{pmatrix} 0 & 0 & 0 & 0 \\ 0 & 2 & -1 & 0 \\ 0 & 0 & \langle 1 \rangle & 0 \\ 0 & 0 & 0 & 0 \end{pmatrix}_{4,2}^{w5}$$

gives $COR(f,g)(0,4) = 2$.

$$g_{4,0} = \begin{pmatrix} 0 & 0 & 0 & 0 \\ 0 & \langle 2 \rangle & 0 & 0 \\ 3 & 0 & 0 & 0 \\ 0 & 0 & 0 & 0 \end{pmatrix}_{4,2}^{w5} \quad and \quad f = \begin{pmatrix} 0 & 0 & 0 & 0 \\ 0 & \langle 2 \rangle & -1 & 0 \\ 0 & 0 & 1 & 0 \\ 0 & 0 & 0 & 0 \end{pmatrix}_{4,2}^{w5}$$

gives $COR(f,g)(4,0) = 4$. *Thus,*

$$COR(f,g) = \begin{pmatrix} 2 & 0 & 0 & 0 & 0 \\ 0 & 0 & 0 & 0 & 0 \\ 0 & 0 & 0 & 0 & 0 \\ 6 & -3 & 0 & 0 & 0 \\ -2 & 3 & 0 & 0 & 4 \end{pmatrix}_{0,4}^{w5}$$

Parallel implementation of correlation can be given just like in the digital signal situation. Of course, translation and the 180° rotation must be performed in a wraparound mode for signals in $\Re^{Z_n \times Z_n}$. The procedure is analogous to wraparound convolution with the exception of the $NINETY^2$ operation performed on g prior to its input into the co-zero and range blocks.

Example 6.30 *We will find the correlation of*

$$f = \begin{pmatrix} 2 & -1 \\ 0 & 1 \end{pmatrix}_{0,1}^{w5}$$

and

$$g = \begin{pmatrix} 0 & 2 \\ 3 & 0 \end{pmatrix}_{0,1}^{w5}$$

using the parallel algorithm. The procedure is shown in Figure 6.1. Here,

$$NINETY^2(g) = \tilde{g}$$

is first found and this is input into the co-zero and range blocks, thus

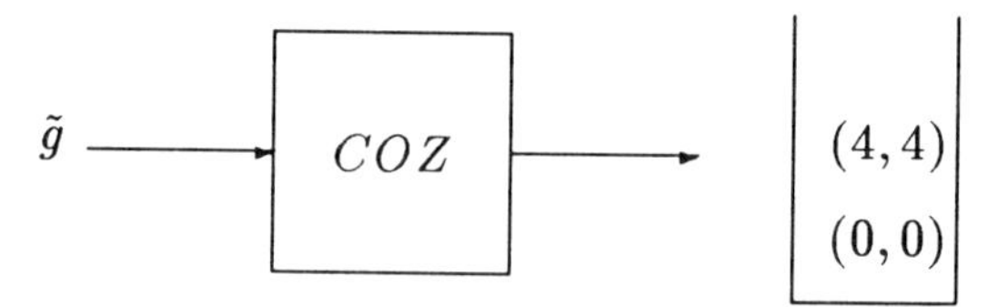

and

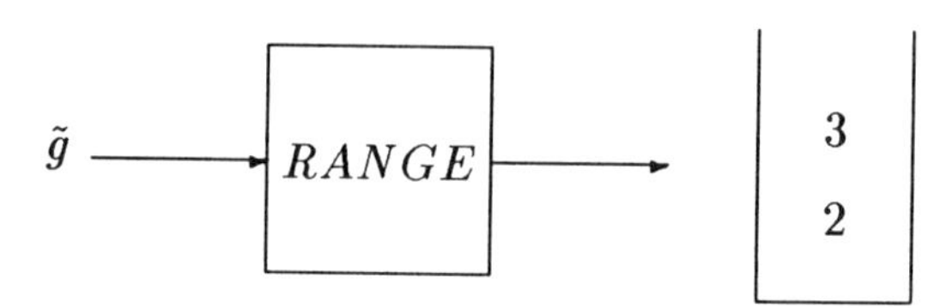

The computation proceeds in the following figure:

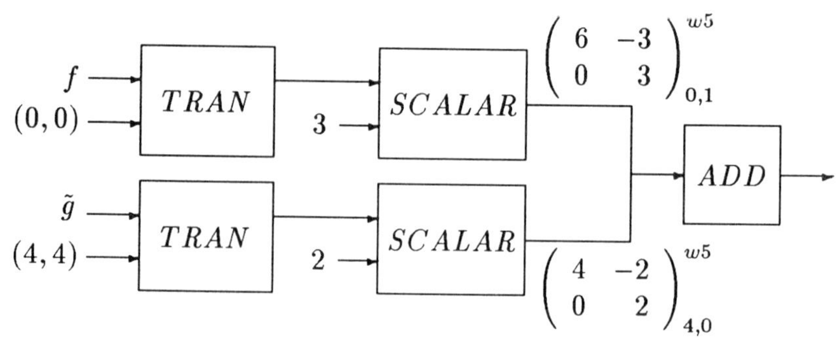

Figure 6.1

Thus we have

$$3f_{0,0} = \begin{pmatrix} 6 & -3 \\ 0 & 3 \end{pmatrix}_{0,1}^{w5}$$

and

$$2f_{4,4} = \begin{pmatrix} 4 & -2 \\ 0 & 2 \end{pmatrix}_{4,0}^{w5}$$

Adding gives the desired result:

$$COR(f,g) = \begin{pmatrix} 2 & 0 & 0 & 0 & 0 \\ 0 & 0 & 0 & 0 & 0 \\ 0 & 0 & 0 & 0 & 0 \\ 6 & -3 & 0 & 0 & 0 \\ -2 & 3 & 0 & 0 & 4 \end{pmatrix}_{0,4}^{w5}$$

All the properties of correlation given in Sections 3.8 and 3.9 hold true for wraparound correlation. Specifically, we have

$$COR(f,g) = NINETY^2(COR(g,f))$$

Also

$$COR(f,g) = CONV(f, NINETY^2(g))$$

and

$$COR(f,g) = NINETY^2(CONV(NINETY^2(f),g))$$

Wraparound correlation is used exactly for the same applications as digital correlation. Additionally, the results given in Section 3.9 for digital correlation also hold for wraparound correlation.

6.7 Wraparound $\mathcal{Z}$ Transforms

The $\mathcal{Z}$ transform for wraparound signals f in $\Re^{Z_n \times Z_n}$ are especially easy to describe. Here

$$f = \begin{pmatrix} f(0,n-1) & \cdots & f(n-1,n-1) \\ \cdot & & \cdot \\ \cdot & \cdots & \cdot \\ \cdot & & \cdot \\ f(0,0) & \cdots & f(n-1,0) \end{pmatrix}_{0,n-1}^{wn}$$

has a $\mathcal{Z}$ transform denoted by $\mathcal{Z}(f)$, or by $F(z, w)$ where

$$F(z, w) = \sum_{k,j=0}^{n-1} f(k, j)\, z^{-k} w^{-j}$$

This relationship is also denoted by

$$f \longleftrightarrow F(z, w)$$

Similar to the usual $\mathcal{Z}$ transforms defined in Section 4.1, four basic operations can be defined. These are: addition, scalar multiplication, multiplication by $z^k w^j$, and inversion. Each of these operations will now be defined in order. Suppose that

$$g = \begin{pmatrix} g(0, n-1) & \cdots & g(n-1, n-1) \\ \cdot & & \cdot \\ \cdot & \cdots & \cdot \\ \cdot & & \cdot \\ g(0, 0) & \cdots & g(n-1, 0) \end{pmatrix}_{0,n-1}^{wn}$$

then

$$\mathcal{Z}(g) = \sum_{k,j=0}^{n-1} g(k, j)\, z^{-k} w^{-j}$$

To find the addition of two $\mathcal{Z}$ transforms, we just add the coefficients of similar powers of z together:

$$F(z, w) + G(z, w) = \sum_{k,j=0}^{n-1} (f(k, j) + g(k, j))\, z^{-k} w^{-j}$$

To multiply the $\mathcal{Z}$ transforms $F(z, w)$ by a real number, a, we multiply each coefficient of $F(z, w)$ by a, thus

$$aF(z, w) = \sum_{k,j=0}^{n-1} af(k, j)\, z^{-k} w^{-j}$$

We form $z^p w^q F(z)$, for integers p and q by first writing

$$z^p w^q F(z, w) = \sum_{k,j=0}^{n-1} f(k, j)\, z^{p-k} w^{q-j}$$

Next, we replace each exponent $i = p - k$ of z by the number r, lying in the set

$$\{0, -1, -2, ..., 1 - n\}$$

obtained by writing

$$i = ns + r \quad \text{where } s \text{ is an integer.}$$

In other words, we used modulo n arithmetic. We do the same for the exponent $q - j$ of w. The end result is that $z^p w^q F(z, w)$ is itself a $\mathcal{Z}$ transform of some signal in $\Re^{Z_n \times Z_n}$. Thus, for instance, if $n = 9$ and the exponent of z is $i = 23$, then we replace i by $r = -4$ since

$$23 = (9)(3) - 4$$

The last operation to be defined at this time is inversion. For $\mathcal{Z}(f)$ it can be found by first writing

$$F(z^{-1}, w^{-1}) = \sum_{k,j=0}^{n-1} f(k, j)\, z^k w^j$$

and then use the modulo n algorithm just described on the exponents of z and w. If this is done then the exponent 0 is left, however, the exponent 1 is replaced by $1 - n$ because

$$1 = n \cdot 1 + (1 - n)$$

Similarly, 2 is replaced by $2 - n$. In general we have $k > 0$ replaced by $n - k$ since

$$\begin{aligned} 2 &= n \cdot 1 + (2 - n) \\ &\vdots \\ n - 2 &= n \cdot 1 + (-2) \\ n - 1 &= n \cdot 1 + (-1) \end{aligned}$$

and so the solution is

$$F\left(z^{-1}, w^{-1}\right) = f\left(0, 0\right) + f\left(0, 1\right) w^{1-n} + f\left(1, 0\right) z^{1-n} + \sum_{k,j=1}^{n-1} f(k, j)\, z^{k-n} w^{j-n}$$

Of consequence, is that the following four structure preserving properties hold true:

Z1) Additive Property: $\mathcal{Z}(f+g) = \mathcal{Z}(f) + \mathcal{Z}(g)$
Z2) Homogeneous Property: $a\mathcal{Z}(f) = \mathcal{Z}(af)$
Z3) Time Shifting Property: $\mathcal{Z}(T(f;n,m)) = z^{-n}w^{-m}F(z,w)$
Z4) Time Reversal Property: $\mathcal{Z}(N^2(f)) = F(1/z,1/w)$

Example 6.31 *Let*

$$f = \begin{pmatrix} 0 & 0 & 0 & 0 \\ 0 & 0 & 2 & 0 \\ 0 & -1 & 3 & 0 \\ 0 & 0 & 0 & 0 \end{pmatrix}_{0,3}^{w4}$$

and

$$g = \begin{pmatrix} 0 & 0 & 1 & 0 \\ 0 & 0 & -1 & 1 \\ 0 & 0 & 2 & 0 \\ 0 & 0 & 0 & 0 \end{pmatrix}_{0,3}^{w4}$$

then

$$f+g = \begin{pmatrix} 0 & 0 & 1 & 0 \\ 0 & 0 & 1 & 1 \\ 0 & -1 & 5 & 0 \\ 0 & 0 & 0 & 0 \end{pmatrix}_{0,3}^{w4}$$

and

$$2f = \begin{pmatrix} 0 & 0 & 0 & 0 \\ 0 & 0 & 4 & 0 \\ 0 & -2 & 6 & 0 \\ 0 & 0 & 0 & 0 \end{pmatrix}_{0,3}^{w4}$$

and

$$T(f;2,2) = \begin{pmatrix} 3 & 0 & 0 & -1 \\ 0 & 0 & 0 & 0 \\ 0 & 0 & 0 & 0 \\ 2 & 0 & 0 & 0 \end{pmatrix}_{0,3}^{w4}$$

and finally,

$$N^2(f) = \begin{pmatrix} 0 & 0 & 3 & -1 \\ 0 & 0 & 2 & 0 \\ 0 & 0 & 0 & 0 \\ 0 & 0 & 0 & 0 \end{pmatrix}_{0,3}^{w4}$$

Next,

$$f \longleftrightarrow F(z,w) = -z^{-1}w^{-1} + 2z^{-2}w^{-2} + 3z^{-2}w^{-1}$$

and

$$g \longleftrightarrow G(z,w) = z^{-2}w^{-3} - z^{-2}w^{-2} + z^{-3}w^{-2} + 2z^{-2}w^{-1}$$

and so

$$F(z,w)+G(z,w) = -z^{-1}w^{-1}+z^{-2}w^{-3}+z^{-2}w^{-2}+5z^{-2}w^{-1}+z^{-3}w^{-2}$$

Also,

$$2F(z,w) = -2z^{-1}w^{-1} + 4z^{-2}w^{-2} + 6z^{-2}w^{-1}$$

and

$$z^{-2}w^{-2}F(z,w) = -z^{-3}w^{-3} + 2z^{-4}w^{-4} + 3z^{-4}w^{-3}$$

$$= -z^{-3}w^{-3} + 2 + 3w^{-3}$$

and finally

$$F(1/z,1/w) = -zw + 2z^2w^2 + 3z^2w$$

$$= -z^{-3}w^{-3} + 2z^{-2}w^{-2} + 3z^{-2}w^{-3}$$

We have, as properties **Z1**) – **Z4**) *state,*

$$f + g \longleftrightarrow F(z,w) + G(z,w)$$

$$2f \longleftrightarrow 2F(z,w)$$

$$T(f;2,2) \longleftrightarrow z^{-2}F(z,w)$$

$$N^2(f) \longleftrightarrow F(1/z,1/w)$$

Property **Z3**) provides the justification for the block diagram given earlier where $1/z$ denotes a one unit shift to the right and $1/w$ denotes a one shift up.

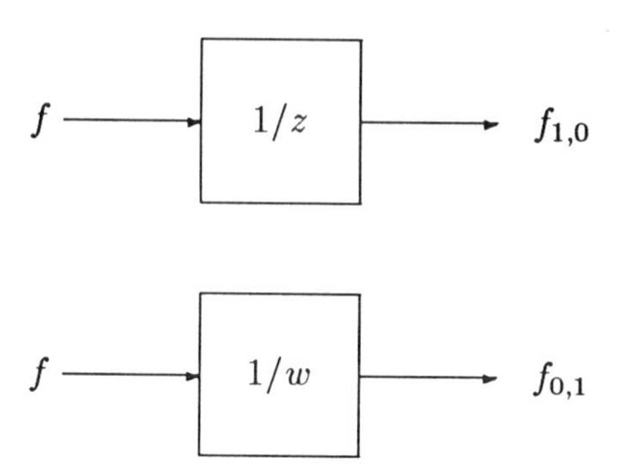

Thus, to shift a signal by k units to the right, we can multiply the $\mathcal{Z}$ transform of the given signal by z^{-k} and form the inverse $\mathcal{Z}$ transform to obtain the result. A similar discussion can be given for translation in the vertical direction.

A most important consequence of the isomorphism given by properties **Z1**) – **Z4**). is that, the convolution of wraparound signals has a $\mathcal{Z}$ transform equal to the product of the $\mathcal{Z}$ transforms of the individual arguments.

Theorem 6.1 *(Wraparound Convolution Theorem) If f and g are in $\Re^{Z_n \times Z_n}$ and*

$$f \longleftrightarrow F(z, w)$$

and

$$g \longleftrightarrow G(z, w)$$

then

$$f * g \longleftrightarrow F(z, w)G(z, w)$$

The product in the above theorem is found using simple algebra along with properties **Z1**) – **Z4**).

Example 6.32 *As in Example 6.27, if*

$$f = \begin{pmatrix} 2 & 1 \\ -1 & 3 \end{pmatrix}_{1,1}^{w4}$$

and

$$g = \begin{pmatrix} 1 & 2 \\ 0 & -1 \end{pmatrix}_{2,1}^{w4}$$

then

$$f \longleftrightarrow F(z,w) = 2z^{-1}w^{-1} + z^{-2}w^{-1} - z^{-1} + 3z^{-2}$$

and

$$g \longleftrightarrow G(z,w) = z^{-2}w^{-1} + 2z^{-3}w^{-1} - z^{-3}$$

and so

$$F(z,w)G(z,w) = 2z^{-3}w^{-2} + 5z^{0}w^{-2} - z^{-3}w^{-1} - z^{0}w^{-1}$$

$$+ 5z^{-1}w^{-1} + z^{0} - 3z^{-1}$$

and so

$$f * g = \begin{pmatrix} 0 & 0 & 0 & 0 \\ 5 & 2 & 0 & 2 \\ -1 & 5 & 0 & -1 \\ 1 & -3 & 0 & 0 \end{pmatrix}_{0,3}^{w4} = \begin{pmatrix} 2 & 5 & 2 \\ -1 & -1 & 5 \\ 0 & 1 & -3 \end{pmatrix}_{3,2}^{w4}$$

as before.

As might be expected, there is a wraparound correlation theorem.

Theorem 6.2 *(Wraparound Correlation Theorem) If f and g are in $\Re^{Z_n \times Z_n}$ and*

$$f \longleftrightarrow F(z,w)$$

and

$$g \longleftrightarrow G(z,w)$$

then

$$COR(f,g) \longleftrightarrow F(z,w)G(1/z,1/w)$$

Example 6.33 *If*

$$f = \begin{pmatrix} 2 & -1 \\ 0 & 1 \end{pmatrix}_{0,1}^{w5}$$

and

$$g = \begin{pmatrix} 0 & 2 \\ 3 & 0 \end{pmatrix}_{0,1}^{w5}$$

Then

$$f \longleftrightarrow F(z,w) = 2w^{-1} - z^{-1}w^{-1} + z^{-1}$$

and

$$g \longleftrightarrow G(z,w) = 2z^{-1}w^{-1} + 3$$

and so

$$N^2(g) \longleftrightarrow G(1/z, 1/w) = 2zw + 3 = 2z^{-4}w^{-4} + 3$$

Now, $F(z,w)G(1/z,1/w)$ *will be found. In doing this, we use the last expression for* $G(1/z,1/w)$. *Thus,*

$$F(z,w)G(1/z,1/w) = 4z^{-4} - 2 + 2w^{-4} + 6w^{-1} - 3z^{-1}w^{-1} + 3z^{-1}$$

and

$$COR(f,g) = \begin{pmatrix} 2 & 0 & 0 & 0 & 0 \\ 0 & 0 & 0 & 0 & 0 \\ 0 & 0 & 0 & 0 & 0 \\ 6 & -3 & 0 & 0 & 0 \\ -2 & 3 & 0 & 0 & 4 \end{pmatrix}_{0,4}^{w5}$$

as was previously obtained in Example 6.28.

6.8 Discrete Fourier Transforms

For any wraparound signal f in $\Re^{Z_n \times Z_n}$ we can find the discrete Fourier transform, abbreviated DFT. It is defined by

$$DFT(f)(r,s) = \sum_{k,i=0}^{n-1} f(k,i)e^{-2\pi j(rk+si)/n}, \text{ where}$$
$$0 \le r,\ s \le n-1 \text{ and } j = \sqrt{-1}$$

The block diagram of the DFT is

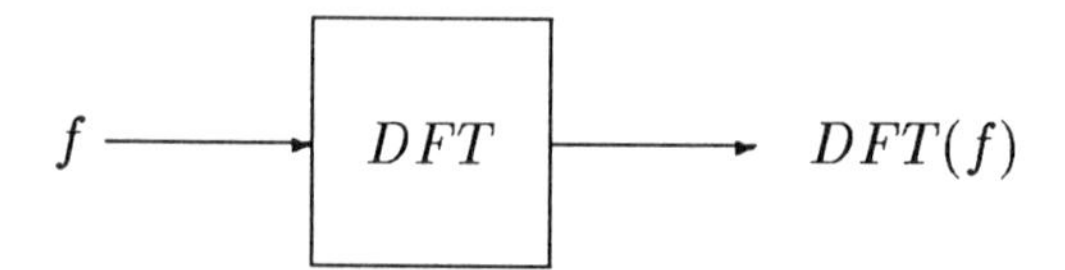

The DFT provides a measure of the frequency content, or spectral content of a signal. The frequency possessed by a signal is indicative of the amount of variation within the signal. While the DFT is useful in filtering signals directly in the "frequency domain" it also provides insight into spatial domain filtering using convolution.

Example 6.34 *Suppose that h is in $\Re^{Z_n \times Z_n}$ where*

$$h = \begin{pmatrix} 0 & 1/5 & 0 \\ 1/5 & 1/5 & 1/5 \\ 0 & 1/5 & 0 \end{pmatrix}_{n-1,1}^{wn}$$

with $n > 2$, in this case. The structuring signal h provides a moving average filter similar to the one given in Section 3.7. Here, h can also be written as

$$h = \begin{pmatrix} 1/5 & 0 & 0 & \cdots & 0 & 0 \\ 0 & 0 & 0 & \cdots & 0 & 0 \\ 0 & 0 & 0 & \cdots & 0 & 0 \\ 1/5 & 0 & 0 & \cdots & 0 & 0 \\ \boxed{1/5} & 1/5 & 0 & \cdots & 0 & 1/5 \end{pmatrix}^{wn}$$

The DFT of h is the complex valued function $DFT(h)$ where

$$DFT(h)(r,s) = \frac{1}{5}\Big[1 + e^{(-2\pi jr)/n} + e^{(-2\pi jr[n-1])/n} + e^{(-2\pi js)/n} + e^{(-2\pi js[n-1])/n}\Big]$$

In general, if $f \in \Re^{Z_n \times Z_n}$ then $DFT(f) \in \mathbf{C}^{Z_n \times Z_n}$, where $\mathbf{C}$ is the set of all complex numbers. The DFT will be a wraparound signal, however its values are always complex valued. It is most helpful to use the amplitude and phase signals associated with the DFT, for they are real. In order to define these quantities in a rigorous fashion, we should introduce a whole new set of range induced operations. This time, the codomain is the set of complex numbers. If this were done we would arrive at many similar operations to those previously found, this time valid for wraparound signals in $\mathbf{C}^{Z_n \times Z_n}$. Define the amplitude spectrum, denoted by $\mid f \mid$, by

$$\mid f \mid (k,i) = \mid f(k,i) \mid$$

Additionally, we also define the real part, imaginary part, and argument type signals for f in $\mathbf{C}^{Z_n \times Z_n}$, denoted by $REAL(f), IMAG(f)$, and $ARG(f)$ respectively. These are defined in turn by

$$REAL(f)(k,i) = Re(f(k,i))$$

where $Re(f(k,i))$ is the real part of the complex number $f(k,i)$,

$$IMAG(f)(k,i) = Im(f(k,i))$$

where $Im(f(k,i))$ is the imaginary part of the complex number $f(k,i)$ and

$$ARG(f)(k,i) = \tan^{-1}\left(\frac{Im(f(k,i))}{Re(f(k,i))}\right)$$

Example 6.35 *Again refer to the wraparound signal*

$$h = \begin{pmatrix} 0 & 1/5 & 0 \\ 1/5 & 1/5 & 1/5 \\ 0 & 1/5 & 0 \end{pmatrix}_{n-1,1}^{wn}$$

in $\Re^{Z_n \times Z_n}$ given in Example 6.34. In this example, we can simplify the DFT by using trigonometric identities to obtain

$$DFT(h)(r,s) = \tfrac{1}{5}\left[1 + 2\cos\left(\tfrac{2\pi r}{n}\right) + 2\cos\left(\tfrac{2\pi s}{n}\right)\right]$$

$$where \quad 0 \leq r,\ s \leq n-1$$

Example 6.36 *Consider the impulse function*

$$\delta_{1,1} = \begin{pmatrix} 0 & 1 \\ 0 & 0 \end{pmatrix}_{0,1}^{wn}$$

Then

$$DFT(\delta_{1,1})(r,s) = e^{-2\pi j[r+s]/n},\ 0 \leq r,\ s \leq n-1$$

So

$$DFT(\delta_{1,1}) = \begin{pmatrix} e^{-2\pi j[n-1]/n} & \cdots & e^{-2\pi j[2n-2]/n} \\ \cdot & & \\ \cdot & & \\ \cdot & & \\ 1 & \cdots & e^{-2\pi j[n-1]/n} \end{pmatrix}_{0,n-1}^{wn}$$

where

$$REAL(\delta_{1,1}) = \begin{pmatrix} \cos(2\pi(n-1))/n & \cdots & \cos(2\pi(2n-2))/n \\ \cdot & & \\ \cdot & & \\ \cdot & & \\ 1 & \cdots & \cos(2\pi(n-1))/n \end{pmatrix}_{0,n-1}^{wn}$$

and

$$IMAG(\delta_{1,1}) = \begin{pmatrix} -\sin(2\pi(n-1))/n & \cdots & -\sin(2\pi(2n-2))/n \\ \cdot & & \\ \cdot & & \\ \cdot & & \\ 1 & \cdots & -\sin(2\pi(n-1))/n \end{pmatrix}_{0,n-1}^{wn}$$

There exists a straight forward relationship between the wraparound $\mathcal{Z}$ transform and the DFT. Indeed, if

$$f \longleftrightarrow F(z,w)$$

then

$$DFT(f)(r,s) = F(e^{-2\pi jr/n}, e^{-2\pi js/n})$$

Several properties possessed by the DFT follow from this. In particular, for f and g in $\Re^{Z_n \times Z_n}$ and a real, we have

$$DFT(f+g) = DFT(f) + DFT(g)$$

$$DFT(af) = aDFT(f)$$

$$DFT(f_{k.i}) = e^{-2\pi jk/n}e^{-2\pi ji/n}DFT(f)$$

$$DFT(CONV(f,g)) = DFT(f)DFT(g)$$

Analogous results hold for f and g in $\mathbf{C}^{Z_n \times Z_n}$. We can define an inverse discrete Fourier transform denoted by $IDFT$ which is defined by

$$IDFT(g)(p,q) = \frac{1}{n^2}\sum_{k,i=0}^{n-1} g(i,k)e^{2\pi jip/n}e^{2\pi jkq/n}$$

for

$$0 \leq p \leq n-1 \text{ and } 0 \leq q \leq n-1$$

The block diagram illustrating this operation is

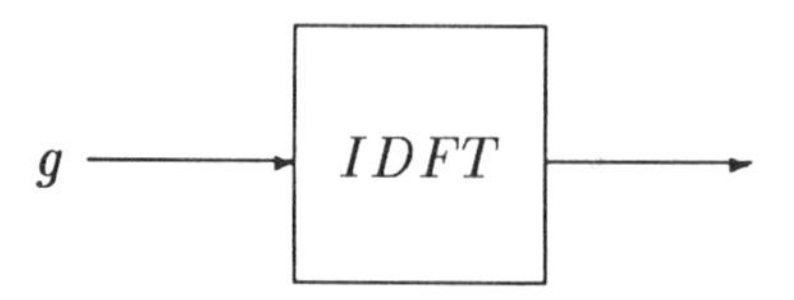

Of particular interest is the fact that for $f, g \in \mathbf{C}^{Z_n \times Z_n}$ we have

$$IDFT(DFT(f)) = f$$

and

$$DFT(IDFT(g)) = g$$

6.9 Wraparound Algebra in $\mathbf{Z}_n \times \mathbf{Z}_m$

In the previous sections we studied wraparound signals in $\Re^{Z_n \times Z_n}$. Several operations on signals in this space also hold for wraparound signals in $\Re^{Z_n \times Z_m}$, where n need not equal m. To begin, all range induced operations: $+$, $\cdot$, and $\vee$, along with range induced macros are defined precisely the same way as those given in Section 6.2. For the domain induced operations and corresponding macro operations we only use the shift, translation operation, and 180° rotation. In particular, the last two operations are denoted by $TRAN$ and $NINETY^2$, or by T and N^2. These are defined by:

$$TRAN(f; p, q)(i, j) = f(i-p, j-q)$$

and

$$NINETY^2(f)(i, j) = f(-i, -j)$$

where $i - p$ is evaluated modulo n and $j - q$ is evaluated modulo m. Similarly, $-i$ is evaluated modulo n, and $-j$ is evaluated modulo m.

The ninety degree rotation N and reflection D operations will not be defined in this space since these operations are not 1-1 or onto mappings.

Example 6.37 *Consider the wraparound signals f and g in $\Re^{Z_3 \times Z_4}$, where*

$$f = \begin{cases} 2 & (0,0) \\ 3 & (1,1) \\ 4 & (0,1) \\ 0 & elsewhere \end{cases}$$

and

$$g = \begin{cases} -1 & (0,0) \\ 2 & (1,0) \\ 1 & (0,1) \\ 0 & elsewhere \end{cases}$$

Bound matrices can also be used to represent f and g, as was the case for wraparound signals defined in a square domain. Here we use bound matrices with 4 rows and 3 columns and notation quite similar to before.

$$f = \begin{pmatrix} 0 & 0 & 0 \\ 0 & 0 & 0 \\ 4 & 3 & 0 \\ 2 & 0 & 0 \end{pmatrix}_{0,3}^{w3\times 4} = \begin{pmatrix} 4 & 3 \\ 2 & 0 \end{pmatrix}_{0,1}^{w3\times 4}$$

$$g = \begin{pmatrix} 0 & 0 & 0 \\ 0 & 0 & 0 \\ 1 & 0 & 0 \\ -1 & 2 & 0 \end{pmatrix}_{0,3}^{w3\times 4} = \begin{pmatrix} 1 & 0 \\ -1 & 2 \end{pmatrix}_{0,1}^{w3\times 4}$$

$$ADD(f,g) = f + g = \begin{pmatrix} 0 & 0 & 0 \\ 0 & 0 & 0 \\ 5 & 0 & 0 \\ 1 & 2 & 0 \end{pmatrix}_{0,3}^{w3\times 4}$$

$$MULT(f,g) = f \cdot g = \begin{pmatrix} 0 & 0 & 0 \\ 0 & 0 & 0 \\ 4 & 0 & 0 \\ -2 & 0 & 0 \end{pmatrix}_{0,3}^{w3\times 4}$$

$$MAX(f,g) = f \vee g = \begin{pmatrix} 0 & 0 & 0 \\ 0 & 0 & 0 \\ 4 & 3 & 0 \\ 2 & 2 & 0 \end{pmatrix}_{0,3}^{w3\times 4}$$

$$T(f;2,1) = \begin{pmatrix} 0 & 0 & 0 \\ 0 & 0 & 0 \\ 3 & 0 & 4 \\ 0 & 0 & 2 \end{pmatrix}_{0,3}^{w3\times 4} = \begin{pmatrix} 3 & 0 & 4 \\ 0 & 0 & 2 \end{pmatrix}_{0,1}^{w3\times 4} = \begin{pmatrix} 4 & 3 \\ 2 & 0 \end{pmatrix}_{0,1}^{w3\times 4}$$

$$N^2(f) = \begin{pmatrix} 4 & 0 & 3 \\ 0 & 0 & 0 \\ 0 & 0 & 0 \\ 2 & 0 & 0 \end{pmatrix}_{0,3}^{w3\times 4}$$

In the last expression of Example 6.37, notice that we set

$$f(-i,-j) = f(1,1) = 3$$

Then we use modulo 3 arithmetic for $-i = 1$ or $i = -1$ to obtain $i = 2$. Then we use modulo 4 arithmetic for $-j = 1$ or $j = -1$ to obtain $j = 3$. This gives

$$N(f)(2,3) = 3$$

All the macro operations which involve the addition, multiplication, maximum, translation, and 180° rotation operations also exist in $\Re^{Z_n \times Z_m}$. In particular, we mention the convolution and correlation operation. Both of these operations can be defined in a pointwise manner. Moreover, they can be implemented using the geometric bound matrix approach as well as using parallel algorithms. For convolution of f and g, which is again denoted by $CONV(f,g)$ or by $f * g$, we have

$$(f*g)(p,q) = \sum_{j=0}^{m-1} \sum_{k=0}^{n-1} f(k,j)\, g(p-k, q-j)$$

As before, the value of the convolution product will be zero except for possibly in the dilation of the co-zero sets of f and g. Thus, we obtain:

$$(f*g)(p,q) = \begin{cases} \sum_{\substack{(k,j)\in coz(f) \\ (p-k,q-j)\in coz(g)}} f(k,j)\,g(p-k,q-j) & (p,q) \in \mathcal{D}(coz(f), coz(g)) \\ 0 & \text{otherwise} \end{cases}$$

The parallel convolution algorithm is illustrated in Figure 3.1 and is given by

$$f * g = \sum_{(p,q)\in coz(g)} f_{p,q} \cdot g(p,q)$$

Similarly, the pointwise correlation is given by:

$$COR(f,g)(p,q) =$$

$$\begin{cases} \sum_{\substack{(k,j)\in coz(f) \\ (k-p,j-q)\in coz(g)}} f(k,j)\,g(k-p,j-q) & (p,q) \in \mathcal{D}(coz(f), coz(N^2(g))) \\ 0 & \text{otherwise} \end{cases}$$

The parallel correlation algorithm is illustrated in Figure 3.2 and is given by

$$COR(f,g) = \sum_{(p,q)\in coz(N^2(g))} f_{p,q} N^2(g)(p,q)$$

Example 6.38 *We will find $f * g$ and $COR(f,g)$ where*

$$f = \begin{pmatrix} 4 & 3 \\ 2 & 0 \end{pmatrix}_{0,1}^{w3\times 4}$$

and

$$g = \begin{pmatrix} 1 & 0 \\ -1 & 2 \end{pmatrix}_{0,1}^{w3\times 4}$$

The parallel algorithm is utilized (see Figure 6.2). As before, the wraparound signal g is input into the co-zero and range blocks.

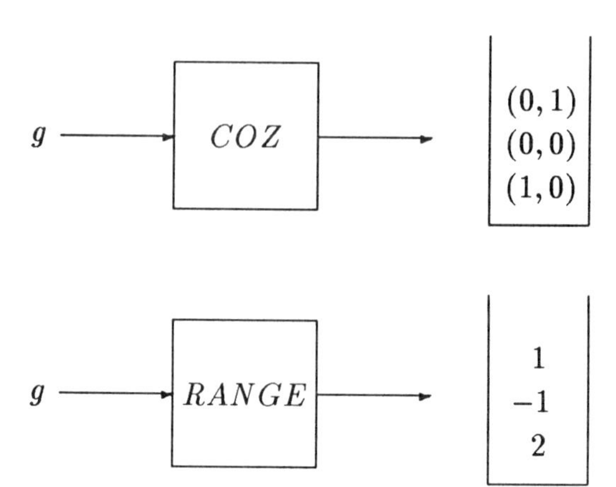

The output of the co-zero and range blocks are sent to the translation and scalar blocks respectively. Finally, the addition is taken of all the translation scalar connections, thereby giving:

$$f * g = \begin{pmatrix} 4 & 3 & 0 \\ -2 & 5 & 6 \\ -2 & 4 & 0 \end{pmatrix}_{0,3}^{w3\times 4}$$

The parallel correlation algorithm is performed in exactly the same fashion, with the exception that $\tilde{g}$ is found first. See the following example and Figure 6.3.

Example 6.39 *Using f and g of Example 6.38,*

$$\tilde{g} = NINETY^2(g) = \begin{pmatrix} 1 & 0 & 0 \\ 0 & 0 & 0 \\ 0 & 0 & 0 \\ -1 & 0 & 2 \end{pmatrix}_{0,3}^{w3\times 4} = \begin{pmatrix} 2 & -1 \\ 0 & 1 \end{pmatrix}_{2,0}^{w3\times 4}$$

This signal is input into the co-zero and range blocks, and the output of these two blocks are used in the translation scalar connections. Finally, the addition block is utilized and

$$COR(f,g) = \begin{pmatrix} 2 & 0 & 0 \\ 0 & 0 & 0 \\ 2 & -3 & 8 \\ 2 & 3 & 4 \end{pmatrix}_{0,3}^{w3\times 4}$$

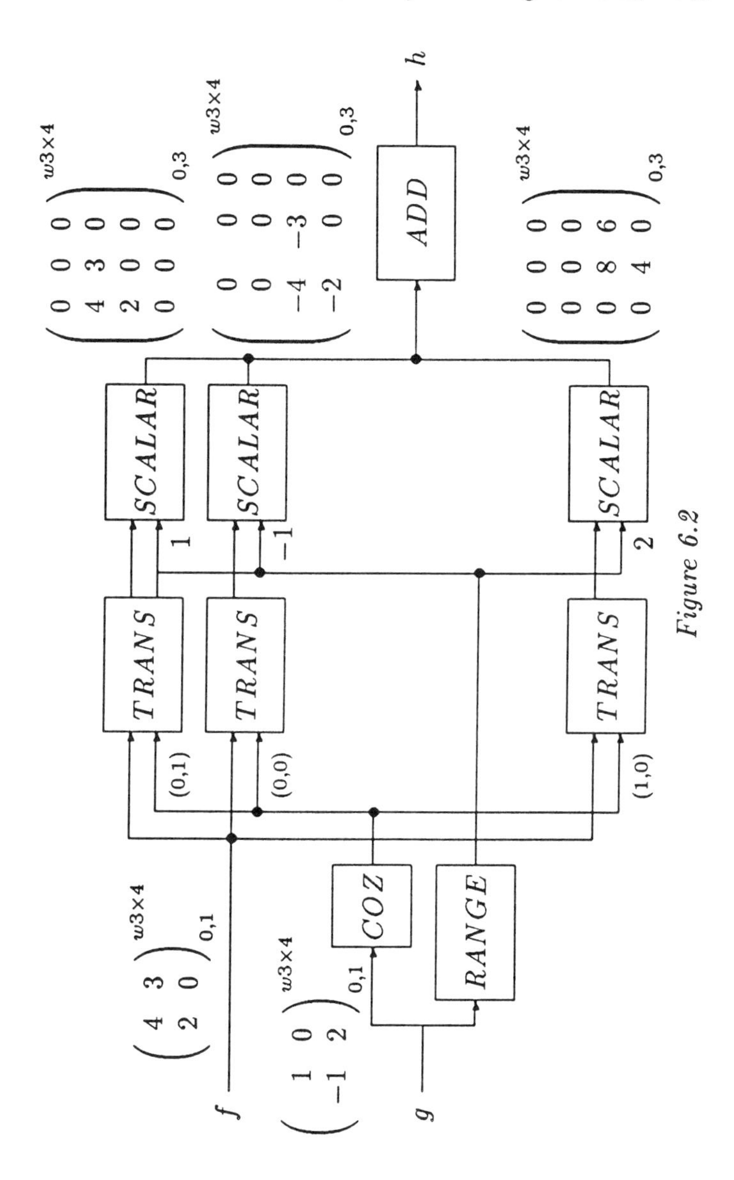
f
g
h
TRANS
TRANS
TRANS
SCALAR
SCALAR
SCALAR
ADD
COZ
RANGE
(0,1)
(0,0)
(1,0)
1
−1
2
w3×4
0,1
0,3

Figure 6.2

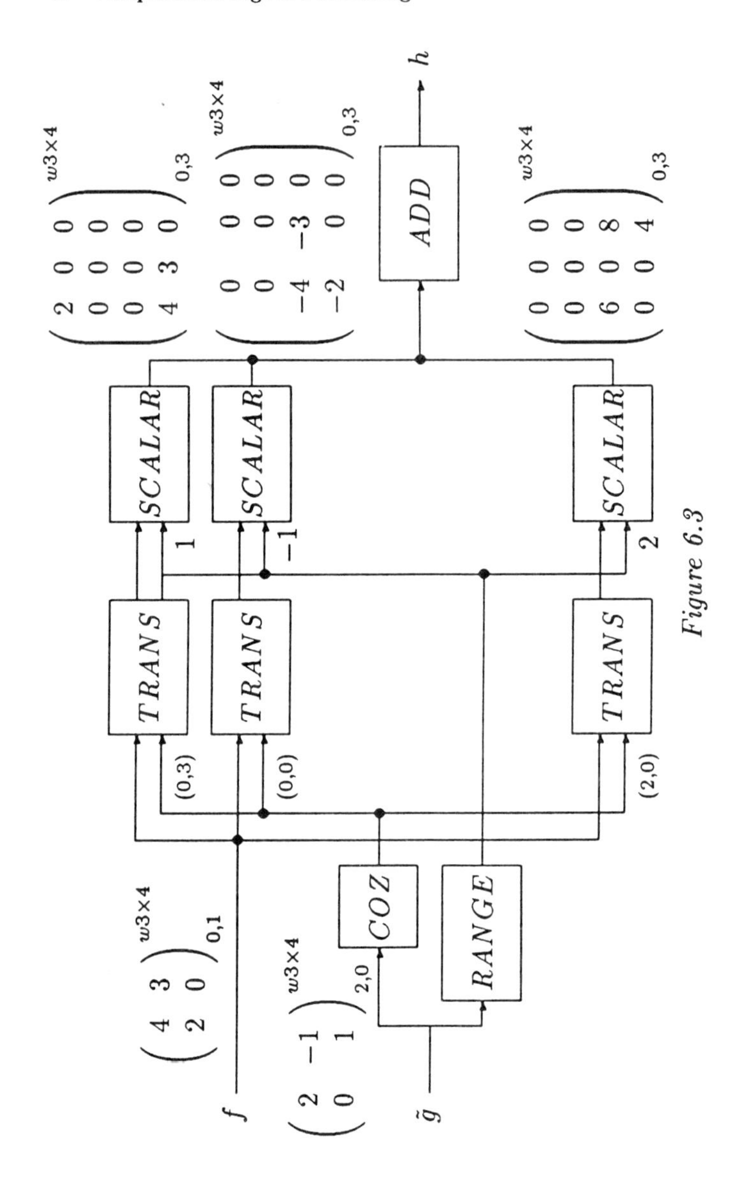

f
g̃
h
COZ
RANGE
TRANS
TRANS
TRANS
SCALAR
SCALAR
SCALAR
ADD
(0,3)
(0,0)
(2,0)
1
−1
2

Figure 6.3

Two dimensional wraparound $\mathcal{Z}$ transforms can be defined for wraparound signals in $\Re^{Z_n \times Z_m}$. This is done just like in $\Re^{Z_n \times Z_n}$. Basic operations involving these series are similar to those given earlier, except that exponents of w are reduced modulo m while exponents of z are reduced modulo n. The principal results involving these transforms are summarized in the following theorem.

Theorem 6.3 *If*

$$f \longleftrightarrow F(z, w)$$

and

$$g \longleftrightarrow G(z, w)$$

then

$$f * g \longleftrightarrow F(z, w)G(z, w)$$

and

$$COR(f, g) \longleftrightarrow F(z, w)G(1/z, 1/w)$$

Example 6.40 *Suppose that*

$$f = \begin{pmatrix} 4 & 3 \\ 2 & 0 \end{pmatrix}_{0,1}^{w3\times4} \quad and \quad g = \begin{pmatrix} 1 & 0 \\ -1 & 2 \end{pmatrix}_{0,1}^{w3\times4}$$

then

$$f \longleftrightarrow F(z, w) = 2 + 4w^{-1} + 3z^{-1}w^{-1}$$

and

$$g \longleftrightarrow G(z, w) = -1 + w^{-1} + 2z^{-1}$$

To begin,

$$F(z, w)G(z, w) = -2 - 2w^{-1} + 5z^{-1}w^{-1} + 4w^{-2}$$

$$+ 3z^{-1}w^{-2} + 4z^{-1} + 6z^{-2}w^{-1}$$

and so by taking the inverse $\mathcal{Z}$ transform we obtain

$$f * g = \begin{pmatrix} 0 & 0 & 0 \\ 4 & 3 & 0 \\ -2 & 5 & 6 \\ -2 & 4 & 0 \end{pmatrix}_{0,3}^{w3\times4}$$

Now, forming

$$G(1/z, 1/w) = -1 + w + 2z$$

and reducing powers of z modulo 3 and powers of w modulo 4 gives

$$G(1/z, 1/w) = -1 + w^{-3} + 2z^{-2}$$

Multiplying

$$F(z, w)G(1/z, 1/w) = 2 - 2w^{-1} - 3z^{-1}w^{-1} + 2w^{-3}$$

$$+ 3z^{-1} + 4z^{-2} + 6z^{-2}w^{-1}$$

Taking the inverse $\mathcal{Z}$ transform gives

$$COR(f, g) = \begin{pmatrix} 2 & 0 & 0 \\ 0 & 0 & 0 \\ 2 & -3 & 8 \\ 2 & 3 & 4 \end{pmatrix}_{0,3}^{w3\times4}$$

The DFT is also defined for signals in $\Re^{Z_n \times Z_m}$. Indeed,

$$DFT(f)(p, q) = \sum_{k=0}^{n-1} \sum_{i=0}^{m-1} f(k, i)e^{-2\pi jkp/n}e^{-2\pi jiq/m}$$

where $0 \le p \le n - 1$ and $0 \le q \le m - 1$

Example 6.41 *Consider the signal $\delta_{1,1}$ in $\Re^{Z_3 \times Z_2}$, then*

$$\delta_{1,1} = \begin{pmatrix} 0 & 1 & 0 \\ 0 & 0 & 0 \end{pmatrix}_{0,1}^{w3\times2}$$

also

$$DFT(f)(p, q) = e^{-2\pi jkp/3}e^{-2\pi ijq/2} \quad \textit{for } 0 \le p \le 2 \textit{ and } 0 \le q \le 1.$$

Thus, we have

$$DFT(f) = \begin{pmatrix} e^{-2\pi j/2} & e^{-10\pi j/6} & e^{-14\pi j/6} \\ 1 & e^{-2\pi j/3} & e^{-4\pi j/3} \end{pmatrix}_{0,1}^{w3\times2}$$

Various properties involving the DFT hold as before and the same is true for the inverse discrete Fourier transform.

6.10 Exercises

1. Provide an illustration of each of the wraparound signals given below.

   (a) $2_{Z_5 \times Z_5}$

   (b) $-1_{Z_3 \times Z_3}$

   (c) δ in $\Re^{Z_4 \times Z_4}$

2. Show that the signals in Exercise 1 can be represented using bound matrices as

   (a)
$$2_{Z_5 \times Z_5} = \begin{pmatrix} 2 & 2 & 2 & 2 & 2 \\ 2 & 2 & 2 & 2 & 2 \\ 2 & 2 & 2 & 2 & 2 \\ 2 & 2 & 2 & 2 & 2 \\ 2 & 2 & 2 & 2 & 2 \end{pmatrix}_{0,4}^{w5}$$

   (b)
$$-1_{Z_3 \times Z_3} = \begin{pmatrix} -1 & -1 & -1 \\ -1 & -1 & -1 \\ -1 & -1 & -1 \end{pmatrix}_{0,3}^{w3}$$

   (c)
$$\delta = \begin{pmatrix} 0 & 0 & 0 & 0 \\ 0 & 0 & 0 & 0 \\ 0 & 0 & 0 & 0 \\ 1 & 0 & 0 & 0 \end{pmatrix}_{0,3}^{w4}$$

   Also, find the co-zero set for each of the wraparound signals given in 2a, 2b, and 2c.

3. Which of the following identities are true:

   (a) $SCALAR(ABS(f); -2) = ABS(SCALAR(f; -2))$

   (b) $OFFSET(ABS(f); -2) = ABS(OFFSET(f; -2))$

   (c) $MULT(ABS(f), ABS(g)) = ABS(MULT(f, g))$

4. For each of the wraparound signals f given in Exercise 1, find the output of the block diagram

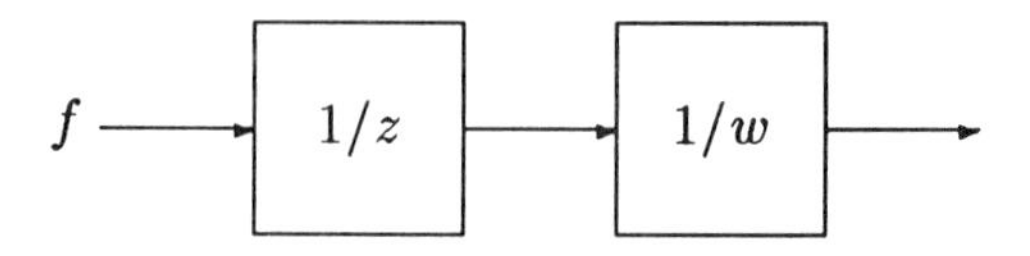

5. Repeat Exercise 4 for the signal f where

$$f = \begin{pmatrix} 2 & 1 & 0 \\ 2 & -1 & 0 \\ 0 & 0 & 0 \end{pmatrix}_{0,2}^{w3}$$

6. For the wraparound signals f given in Exercise 1 determine the output of the block diagram

$$f \longrightarrow \boxed{NINETY^2} \longrightarrow$$

7. Repeat Exercise 6 using

$$f = \begin{pmatrix} 2 & 1 & 0 \\ 0 & 0 & 1 \\ 0 & 0 & 0 \end{pmatrix}_{0,2}^{w3}$$

8. Show that

$$f = \begin{pmatrix} 2 & 0 & 3 & 2 \\ 1 & 0 & 1 & 4 \\ 0 & 0 & 0 & 0 \end{pmatrix}_{0,3}^{w4}$$

can be written as

$$f = \begin{pmatrix} 2 & 0 & 3 & 2 \\ 1 & 0 & 1 & 4 \end{pmatrix}_{0,3}^{w4}$$

as well as

$$f = \begin{pmatrix} 3 & 2 & 2 \\ 1 & 4 & 1 \end{pmatrix}_{2,3}^{w4}$$

What is the most compressed way of writing f?

9. Write
$$f = \begin{pmatrix} 3 & 0 & 0 & 2 \\ 0 & 0 & 0 & 0 \\ 0 & 0 & 0 & 0 \\ 4 & 0 & 0 & -1 \end{pmatrix}_{0,3}^{w4}$$
in the most compressed fashion.

10. Consider the set $A = \{(0,0),(1,1),(1,0)\}$ in $\mathbf{Z}_9 \times \mathbf{Z}_9$ and let $B = \{(1,1),(2,2),(1,0)\}$, find $\mathcal{D}(A,B)$.

11. Find $\mathcal{E}(A,B)$ using A and B given in Exercise 10. Do this using the definition as well as using $A \ominus -B$.

12. Find $\mathcal{O}(A,B)$ using A and B as given in Exercise 10.

13. Find $\mathcal{C}(A,B)$ using A and B as given in Exercise 10.

14. Show by example, using wraparound set morphology, that $N^2(A)$ need not equal $-A$.

15. Let
$$f = \begin{pmatrix} 2 & 1 & 0 \\ 0 & 1 & 0 \\ 0 & 1 & 0 \end{pmatrix}_{0,2}^{w3}$$
and
$$g = \begin{pmatrix} 2 & 0 & 1 \\ 0 & 0 & 0 \\ -1 & 0 & 0 \end{pmatrix}_{0,2}^{w3}$$
Find the output of the block diagram

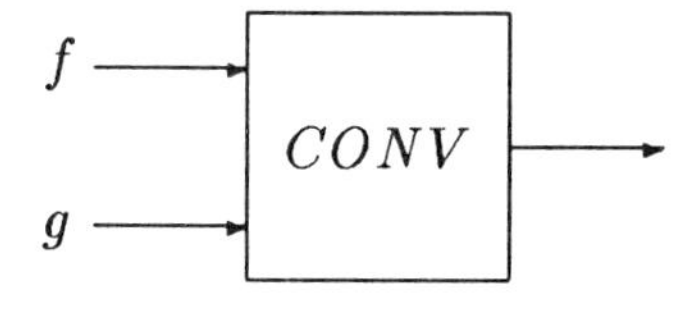

by using

(a) The definition of convolution.

(b) The bound matrix technique.

(c) The parallel convolution algorithm.

16. Use f and g in Exercise 15 and find the output of

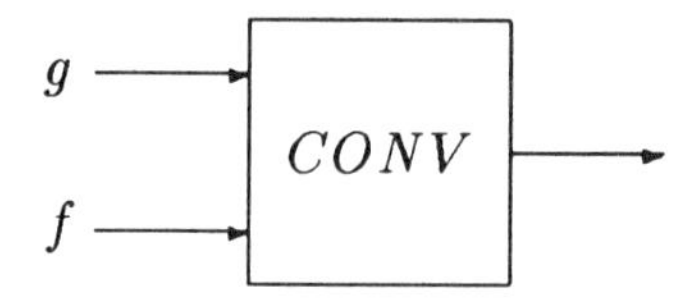

17. Give an example in $\Re^{Z_3 \times Z_3}$ showing that $coz(f * g)$ need not equal $\mathcal{D}(coz(f), coz(g))$.

18. For the wraparound signals f and g given in Exercise 15, find the output of

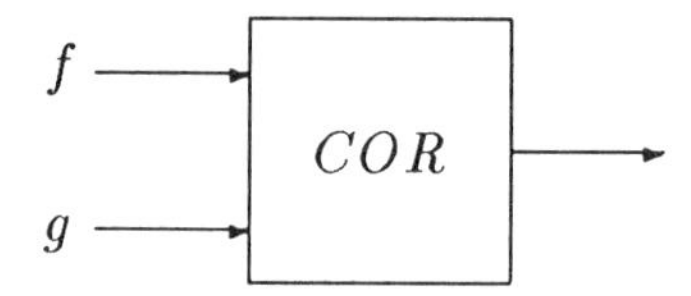

do this by

(a) The definition.

(b) The bound vector technique.

(c) The parallel correlation algorithm.

19. Let

$$f = \begin{pmatrix} 2 & 3 & 4 \\ 1 & 0 & -1 \\ 0 & 0 & 0 \end{pmatrix}_{0,2}^{w3}$$

find the $\mathcal{Z}$ transform for the output of each of the following block diagrams:

(a)

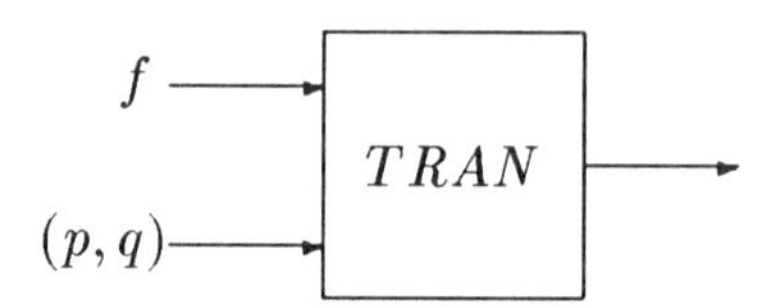

(b)

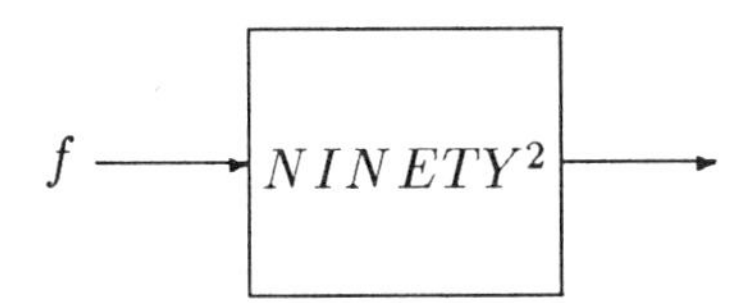

20. Find $f * g$ using $\mathcal{Z}$ transforms where

$$f = \begin{pmatrix} 2 & 3 & 4 \\ 0 & 0 & 1 \\ 0 & 0 & 0 \end{pmatrix}_{0,2}^{w3}$$

and

$$g = \begin{pmatrix} 2 & 1 & 0 \\ -1 & 0 & 0 \\ 0 & 0 & 0 \end{pmatrix}_{0,2}^{w3}$$

21. Use the wraparound signals f and g in Exercise 20 and find $COR(f, g)$ using the $\mathcal{Z}$ transform technique.

22. Find the DFT for $\delta_{1,1}$, do this in

   (a) $\Re^{Z_3 \times Z_1}$

   (b) $\Re^{Z_1 \times Z_3}$

   (c) $\Re^{Z_2 \times Z_4}$

23. Find the minimal bound matrix for

$$f = \begin{pmatrix} 2 & 1 & 0 & 1 \\ 0 & 0 & 0 & 0 \\ 0 & 0 & 0 & 0 \\ 3 & 0 & 0 & 1 \end{pmatrix}_{0,3}^{w3\times 4}$$

24. Let

$$f = \begin{pmatrix} 2 & 1 & 3 \\ 4 & 2 & 0 \end{pmatrix}_{0,1}^{w2\times 3}$$

and

$$g = \begin{pmatrix} -1 & 0 & 1 \\ 3 & 2 & 0 \end{pmatrix}_{0,1}^{w2\times 3}$$

Find $f * g$ by the definition, using bound matrices, and using the parallel algorithms.

25. Find $COR(f, g)$ for f and g given in Exercise 24 by the definition, using bound matrices, and using parallel algorithms.

26. In $\Re^{Z_n \times Z_m}$ with $n \neq m$, show that N is not 1-1 and onto where

$$N(f)(i,j) = f(j,-i)$$

Is $VERT$ and HOR 1-1 and onto? Recall

$$VERT(f)(i,j) = f(-i,j)$$

$$HOR(f)(i,j) = f(i,-j)$$

27. Define an inverse DFT operator $IDFT$ in $\Re^{Z_n \times Z_m}$, and show properties similar to those described in Section 6.8 hold true in this signal algebra environment.

7

Parallel Multidimensional Algorithms for Single Dimensional Signal Processing

7.1 Two Dimensional Processing of One Dimensional Signals

In this section we will describe a simple application of using two dimensional convolution for performing several one dimensional convolutions simultaneously. To do this, we must map one dimensional signals into two dimensional signals. Subsequent sections of this chapter also address the benefits of mapping signals in a given dimension into signals of another dimension.

Suppose that we are given m, one dimensional, time limited signals

$$g^0, g^1, ..., g^{m-1}$$

each of which is to be convolved with a given signal f. We could form a two dimensional signal g, which is represented by a bound matrix with m rows. It has as its bottom row g^0. The next to bottom row will be g^1 and so on, up to its top row which is g^{m-1}. Also, to precisely locate the vectors g^i in g we let

$$g(k,j) = \begin{cases} g^j(k) & j = 0, 1, ..., m-1 \\ 0 & otherwise \end{cases}$$

Next, we convolve g with f where f is represented as a two dimensional signal with values on the abscissa. A two dimensional signal h is obtained which we write as a bound matrix with m rows. The bottom row is h^0, the next to bottom row is h^1, and so on up to the top row, which is h^{m-1}. Specifically,

$$h(k,j) = \begin{cases} h^j(k) & j = 0, 1, ..., m-1 \\ 0 & otherwise \end{cases}$$

It will be true that

$$f * g^i = h^i, \quad i = 0, 1, ..., m-1.$$

Example 7.1 *Say we wish to convolve*

$$g^0 = \begin{pmatrix} 1 & 2 & 2 \end{pmatrix}_0^0$$

and

$$g^1 = \begin{pmatrix} -1 & 2 & 2 \end{pmatrix}_1^0$$

and

$$g^2 = \begin{pmatrix} 3 & 2 & 1 & 1 & 1 \end{pmatrix}_0^0$$

with

$$f = \begin{pmatrix} 2 & 2 & -1 \end{pmatrix}_1^0$$

Form the two dimensional signal

$$g = \begin{pmatrix} 3 & 2 & 1 & 1 & 1 \\ 0 & -1 & 2 & 2 & 0 \\ 1 & 2 & 2 & 0 & 0 \end{pmatrix}_{0,2}^0$$

which will be convolved with

$$f = \begin{pmatrix} 2 & 2 & -1 \end{pmatrix}_{1,0}^0$$

See Figure 7.1, where we obtain

$$h = \begin{pmatrix} 6 & 10 & 3 & 2 & 3 & 1 & -1 \\ 0 & -2 & 2 & 9 & 2 & -2 & 0 \\ 2 & 6 & 7 & 2 & -2 & 0 & 0 \end{pmatrix}_{1,2}^0$$

Thus,

$$f * g^0 = \begin{pmatrix} 2 & 6 & 7 & 2 & -2 \end{pmatrix}_1^0$$

$$f * g^1 = \begin{pmatrix} 0 & -2 & 2 & 9 & 2 & -2 \end{pmatrix}_1^0$$

$$f * g^2 = \begin{pmatrix} 6 & 10 & 3 & 2 & 3 & 1 & -1 \end{pmatrix}_1^0$$

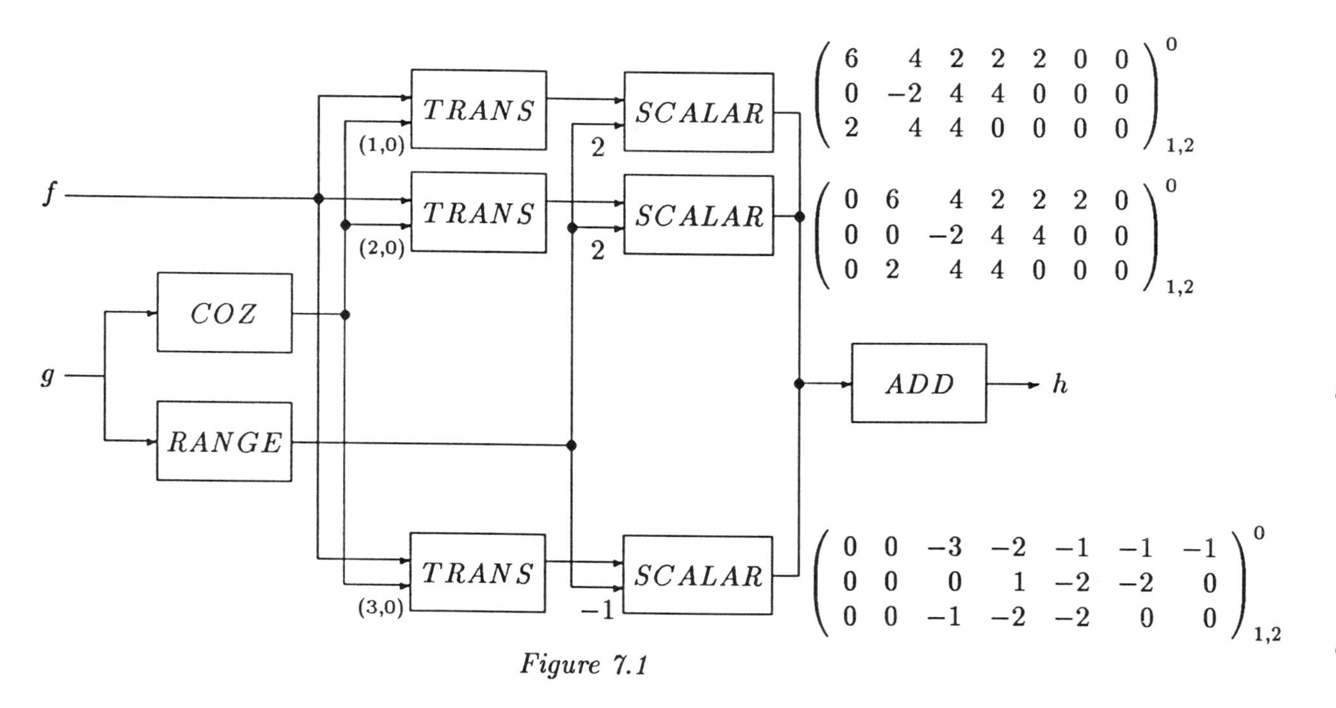
f
g
COZ
RANGE
TRANS
TRANS
TRANS
(1,0)
(2,0)
(3,0)
SCALAR
SCALAR
SCALAR
2
2
−1
ADD
h
6 4 2 2 2 0 0
0 −2 4 4 0 0 0
2 4 4 0 0 0 0
0
1,2
0 6 4 2 2 2 0
0 0 −2 4 4 0 0
0 2 4 4 0 0 0
0
1,2
0 0 −3 −2 −1 −1 −1
0 0 0 1 −2 −2 0
0 0 −1 −2 −2 0 0
0
1,2

Figure 7.1

7.2 Diagonal Transform from One Dimension into Two Dimensions

There are many transformations similar to the one given in the last section, used in taking one dimensional digital signals into several dimensions. Perhaps the simplest is the diagonal transformation, denoted by $DIAG$. Intuitively, it maps a one dimensional digital signal along the 45° line in $\Re^{Z \times Z}$. Specifically, for the one dimensional digital signal f given by the bound vector

$$\left(\cdots \quad a_{-2} \quad a_{-1} \quad \boxed{a_0} \quad a_1 \quad a_2 \quad \cdots \right)$$

we can form the associated diagonal (two dimensional) digital signal $DIAG(f)$ where

$$DIAG(f) = \begin{pmatrix} & \cdot & \cdot & \cdot & \cdot & \cdot & \\ & \cdot & \cdot & \cdot & \cdot & \cdot & \\ & \cdot & \cdot & \cdot & \cdot & \cdot & \\ \cdots & 0 & 0 & 0 & 0 & a_2 & \cdots \\ \cdots & 0 & 0 & 0 & a_1 & 0 & \cdots \\ \cdots & 0 & 0 & a_0 & 0 & 0 & \cdots \\ \cdots & 0 & a_{-1} & 0 & 0 & 0 & \cdots \\ \cdots & a_{-2} & 0 & 0 & 0 & 0 & \cdots \\ & \cdot & \cdot & \cdot & \cdot & \cdot & \\ & \cdot & \cdot & \cdot & \cdot & \cdot & \\ & \cdot & \cdot & \cdot & \cdot & \cdot & \end{pmatrix}$$

Thus,

$$DIAG : \Re^Z \rightarrow \Re^{Z \times Z}$$

where

$$DIAG(f)(n,m) = \begin{cases} f(n) & n = m \\ 0 & otherwise \end{cases}$$

The following block diagram illustrates this operation.

$$f \longrightarrow \boxed{DIAG} \longrightarrow DIAG(f)$$

If f is time limited then $DIAG(f)$ will have finite support. Moreover, if

$$f = \begin{pmatrix} a_0 & a_1 & \cdots & a_{n-1} \end{pmatrix}_p^0$$

then

$$DIAG(f) = \begin{pmatrix} 0 & 0 & 0 & \cdots & a_{n-1} \\ \cdot & \cdot & \cdot & & \cdot \\ \cdot & \cdot & \cdot & & \cdot \\ \cdot & \cdot & \cdot & & \cdot \\ 0 & 0 & a_2 & \cdots & 0 \\ 0 & a_1 & 0 & \cdots & 0 \\ a_0 & 0 & 0 & \cdots & 0 \end{pmatrix}_{p,p+n-1}^0$$

Example 7.2 *Let*

$$f = \begin{pmatrix} 2 & 0 & 3 & 1 & -1 \end{pmatrix}_2^0$$

then

$$DIAG(f) = \begin{pmatrix} 0 & 0 & 0 & 0 & -1 \\ 0 & 0 & 0 & 1 & 0 \\ 0 & 0 & 3 & 0 & 0 \\ 0 & 0 & 0 & 0 & 0 \\ 2 & 0 & 0 & 0 & 0 \end{pmatrix}_{2,6}^0$$

Applications involving this transformation will be given shortly.

Notice that if $DIAG(f)$ is given we can always recover the one dimensional signal f. This is done using the inverse diagonal transform and it is denoted by $IDIAG$. We have, for any two dimensional digital signal g,

$$IDIAG : \Re^{Z \times Z} \rightarrow \Re^Z$$

where for $n \in \mathbf{Z}$

$$IDIAG(g)(n) = g(n, n)$$

In terms of block diagrams

$$g \longrightarrow \boxed{IDIAG} \longrightarrow IDIAG(g)$$

Accordingly, the following property holds true.

D1) Inversion $IDIAG(DIAG(f)) = f$

7.3 Kronecker Products

Another transformation utilized in mapping a one dimensional digital signal into two dimensions is the Kronecker operation, also called Kronecker product. Unlike the diagonal transformation, the Kronecker operation, denoted by $KRON$, or by $\boxed{K}$, is a binary operation. Indeed,

$$KRON : \Re^Z \times \Re^Z \rightarrow \Re^{Z \times Z}$$

For any two one dimensional digital signals f and g, the Kronecker product is denoted by $KRON(f, g)$ or by $f \boxed{K} g$ and it is defined by the following

$$(f \boxed{K} g)(i, j) = f(i)g(j)$$

The block diagram illustrating this operation is given by

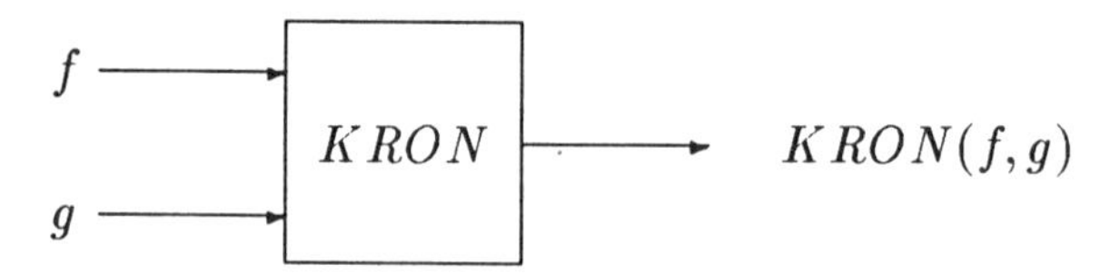

If f and g are both time limited, then $f \boxed{K} g$ will be of finite support. Moreover, if

$$f = \begin{pmatrix} a_0 & a_1 & \cdots & a_{n-1} \end{pmatrix}_p^0$$

and

$$g = \begin{pmatrix} b_0 & b_1 & \cdots & b_{m-1} \end{pmatrix}_q^0$$

then

f
g
KRON

$$\begin{pmatrix} a_0b_{m-1} & a_1b_{m-1} & \cdots & a_{n-1}b_{m-1} \\ \cdot & \cdot & & \cdot \\ \cdot & \cdot & & \cdot \\ \cdot & \cdot & & \cdot \\ a_0b_1 & a_1b_1 & \cdots & a_{n-1}b_1 \\ a_0b_0 & a_1b_0 & \cdots & a_{n-1}b_0 \end{pmatrix}_{p,q+m-1}^0$$

Example 7.3 *Suppose that*

$$f = \begin{pmatrix} 2 & -1 & 3 \end{pmatrix}_3^0 \quad and \quad g = \begin{pmatrix} 2 & 3 \end{pmatrix}_4^0$$

then

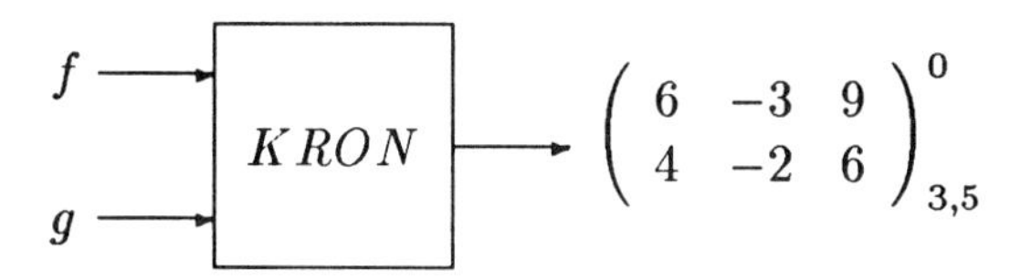

Also, note that

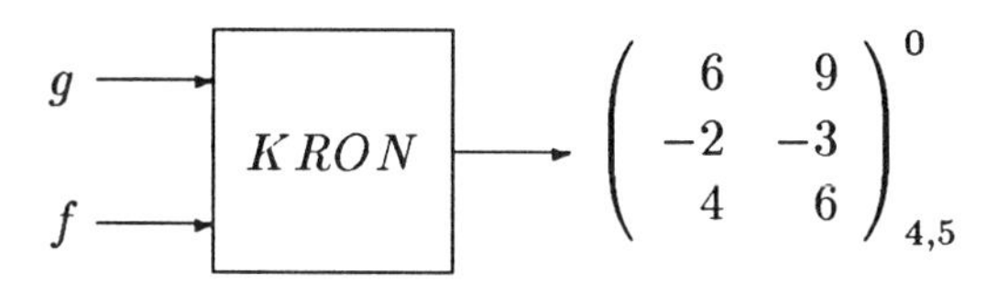

See Figures 7.2 *a* and *b*

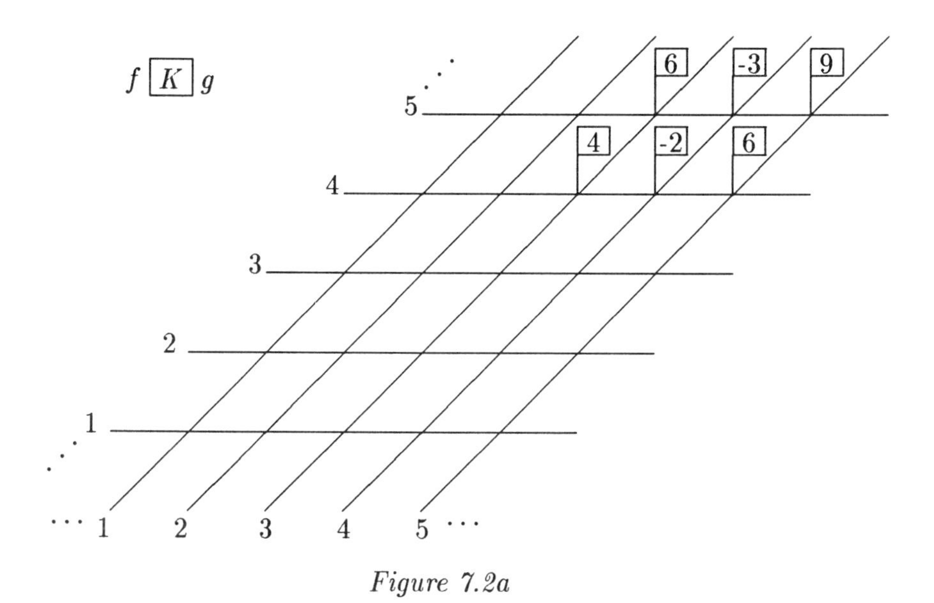

Figure 7.2a

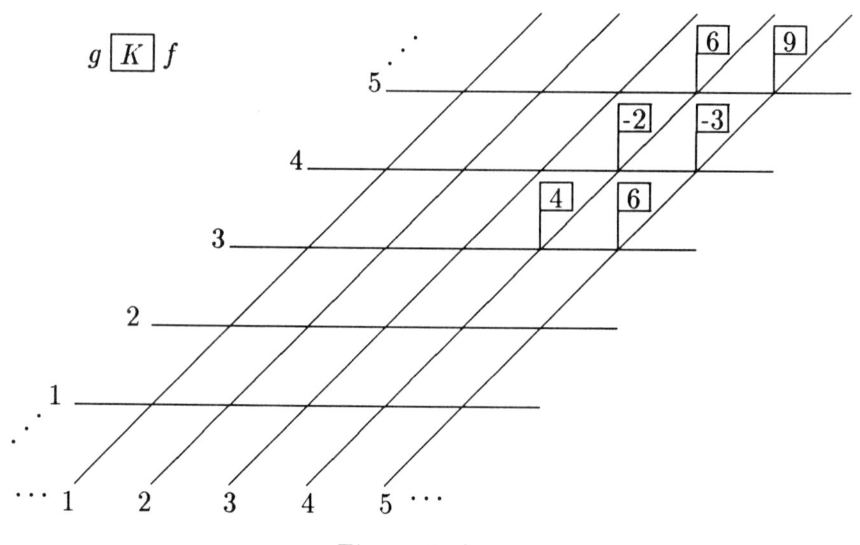

Figure 7.2b

As the above example illustrates, we usually have

$$f \boxed{K} g \neq g \boxed{K} f.$$

More specifically, the following property always holds true for the Kronecker product of two signals:

K1) Diagonally Commutative: $DIFLIP(KRON(f,g)) = KRON(g,f)$

Property **K1)** means that

$$(f \boxed{K} g)(i,j) = (g \boxed{K} f)(j,i)$$

and in terms of block diagrams we have the output of

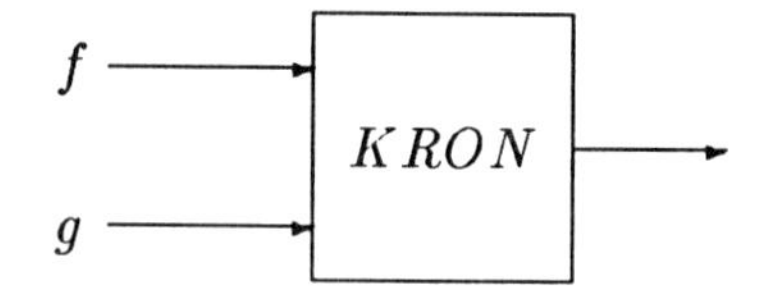

and the output of

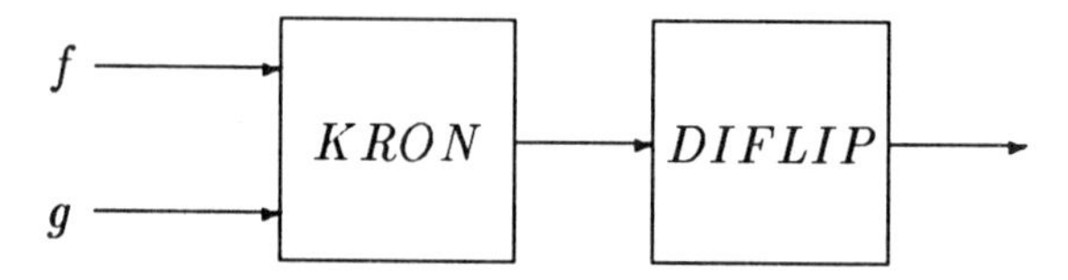

are equal.

The Kronecker product provides a useful way of developing two dimensional structuring signals with properties similar to those in the one dimensional case. Suppose a one dimensional structuring signal (impulse response) is h. The product $h \boxed{K} h$, will be a two dimensional structuring signal, which when used as an argument in convolution, filters in a manner similar to the way that h filters in one dimension.

Example 7.4 *Consider the triangular structuring signal h illustrated in Figure 7.3. This function is used in one dimension as a low pass filter.*

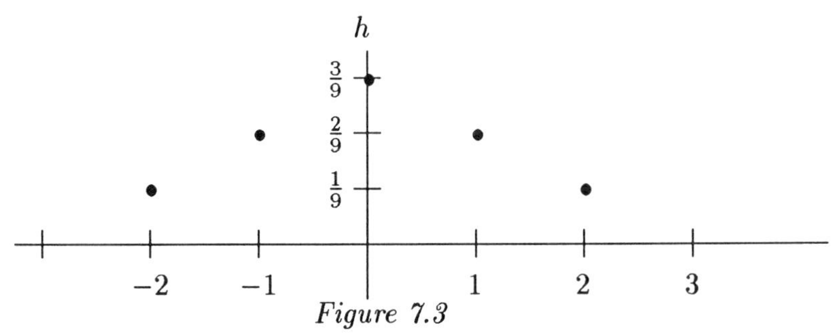

Figure 7.3

Since

$$h = \frac{1}{9}\begin{pmatrix} 1 & 2 & 3 & 2 & 1 \end{pmatrix}_{-2}^{0}$$

we have in two dimensions

$$h \boxed{K} h = \frac{1}{81}\begin{pmatrix} 1 & 2 & 3 & 2 & 1 \\ 2 & 4 & 6 & 4 & 2 \\ 3 & 6 & 9 & 6 & 3 \\ 2 & 4 & 6 & 4 & 2 \\ 1 & 2 & 3 & 2 & 1 \end{pmatrix}_{-2,2}^{0}$$

This signal is also a low pass filter.

Notice that there cannot be an inverse mapping for the Kronecker product. In other words, given $f \boxed{K} g$ there is no way, in general, of uniquely determining both f and g.

Example 7.5 *Let*

$$f = \begin{pmatrix} 1 & 2 \end{pmatrix}_{0}^{0}$$

and

$$g = \begin{pmatrix} 3 & 4 \end{pmatrix}_0^0$$

then

$$f \boxed{K} g = \begin{pmatrix} 4 & 8 \\ 3 & 6 \end{pmatrix}_{0,1}^{0}$$

However, if we use

$$-f = \begin{pmatrix} -1 & -2 \end{pmatrix}_0^0$$

and

$$-g = \begin{pmatrix} -3 & -4 \end{pmatrix}_0^0$$

then

$$-f \boxed{K} -g = \begin{pmatrix} 4 & 8 \\ 3 & 6 \end{pmatrix}_{0,1}^{0}$$

The previous example illustrates

$$f \boxed{K} g = -f \boxed{K} -g$$

and this equality always holds true. More generally, we have the property that for any reals a and b, and any f and g in $\Re^Z$:

K2) Scaling: $af \boxed{K} bg = ab(f \boxed{K} g)$

In terms of block diagrams, the output of the following two block diagrams are always equal:

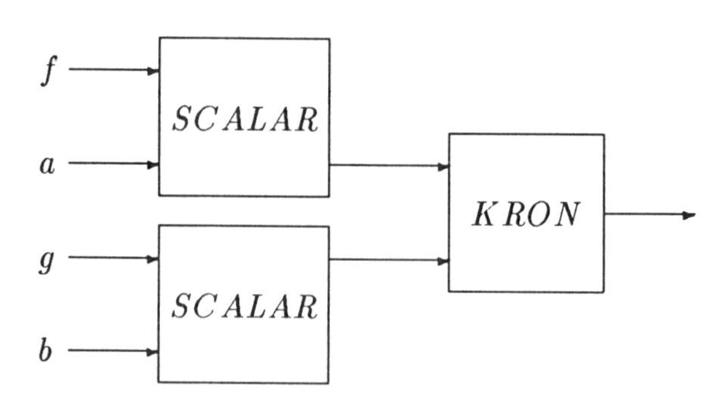

and of

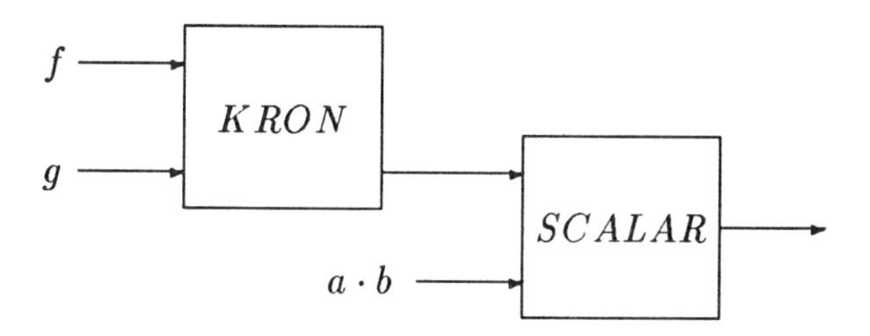

7.4 Triple Convolution

Triple convolution of three one dimensional digital signals f, g and h is formally given by the infinite series

$$\sum_{k=-\infty}^{\infty} f(k)g(n-k)h(m-k)$$

The resulting function, provided that the series converges for all n and m, will be a two dimensional digital signal which we shall denote by $TCONV(f, g, h)$. It has been seen that $DIAG$ is a unary operator from $\Re^Z$ into $\Re^{Z\times Z}$. $TCONV$ is a ternary operation. The only trouble is that

$$TCONV : \Re^Z \times \Re^Z \times \Re^Z \longrightarrow \Re^{Z\times Z}$$

is not true, because the series does not converge for arbitrary functions in $\Re^Z$. This is just like the situation which arises in convolution.

Let S be the subset of $\Re^Z$ consisting of time limited signals, then it is true that

$$TCONV : S \times S \times S \longrightarrow \Re^{Z\times Z}$$

It is this situation which will be described herein. Specifically, the triple convolution will be zero everywhere, except possibly for lattice points (n, m) for which

$$(n, m) \in \mathcal{D}(A, B)$$

where $\mathcal{D}$ is the set dilation operation and where

$$A = \text{coz}(DIAG(f))$$

and

$$B = \operatorname{coz}(KRON(g,h))$$

The triple convolution is defined by

$$TCONV(f,g,h)(n,m) = \begin{cases} \sum\limits_{\substack{k \in \operatorname{coz}(f) \\ n-k \in \operatorname{coz}(g) \\ m-k \in \operatorname{coz}(h)}} f(k)g(n-k)h(m-k) & \begin{matrix}(n,m) \in \\ \mathcal{D}(A,B)\end{matrix} \\ 0 & \text{otherwise} \end{cases}$$

Example 7.6 *The triple convolution of f, g and h is to be found, where*

$$f = \begin{pmatrix} 1 & 2 \end{pmatrix}_0^0$$

$$g = \begin{pmatrix} 1 & 0 & 2 \end{pmatrix}_3^0$$

and

$$h = \begin{pmatrix} 2 & -1 \end{pmatrix}_2^0$$

Since

$$DIAG(f) = \begin{pmatrix} 0 & 2 \\ 1 & 0 \end{pmatrix}_{0,1}^0$$

and

$$KRON(g,h) = \begin{pmatrix} -1 & 0 & -2 \\ 2 & 0 & 4 \end{pmatrix}_{3,3}^0$$

we have

$$A = \operatorname{coz}(DIAG(f)) = \{(0,0),(1,1)\}$$

and

$$B = \operatorname{coz}(KRON(g,h)) = \{(3,3),(3,2),(5,3),(5,2)\}$$

Accordingly, the dilation of these two sets is

$$\mathcal{D}(A,B) = \{(3,3),(3,2),(5,3),(5,2),(4,4),(4,3),(6,4),(6,3)\}$$

We have

$$\begin{aligned} \operatorname{coz}(f) &= \{0,1\} \\ \operatorname{coz}(g) &= \{3,5\} \\ \operatorname{coz}(h) &= \{2,3\} \end{aligned}$$

Thus,

$$TCONV(f,g,h)(3,3) = \sum_{\substack{k\in\{0,1\}\\ n-k\in\{3,5\}\\ m-k\in\{2,3\}}} f(k)g(3-k)h(3-k)$$

$$= f(0)g(3)h(3) = -1$$

Similarly,

$$\begin{array}{rcr} TCONV(f,g,h)(3,2) = f(0)g(3)h(2) = & 2 \\ TCONV(f,g,h)(5,3) = f(0)g(5)h(2) = & -2 \\ TCONV(f,g,h)(5,2) = f(0)g(5)h(2) = & 4 \\ TCONV(f,g,h)(4,4) = f(1)g(3)h(3) = & -2 \\ TCONV(f,g,h)(4,3) = f(1)g(3)h(2) = & 4 \\ TCONV(f,g,h)(6,4) = f(1)g(5)h(3) = & -4 \\ TCONV(f,g,h)(6,3) = f(1)g(5)h(2) = & 8 \end{array}$$

Hence the resulting two dimensional signal is

$$TCONV(f,g,h) = \begin{pmatrix} 0 & -2 & 0 & -4 \\ -1 & 4 & -2 & 8 \\ 2 & 0 & 4 & 0 \end{pmatrix}_{3,4}^{0}$$

Triple convolution can be illustrated using the following block diagram:

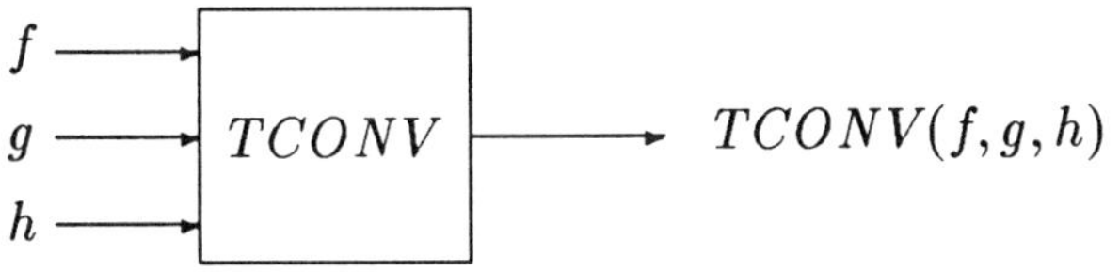

$$\begin{array}{l} f \longrightarrow \\ g \longrightarrow \\ h \longrightarrow \end{array} \boxed{TCONV} \longrightarrow TCONV(f,g,h)$$

7.5 Triple Convolution Performed Using Parallel Two Dimensional Convolution

In this section, we will show how to perform triple convolution using two dimensional convolution. A good indication how this can be done was hinted from the previous section. Indeed, there we saw that a support region for $TCONV(f,g,h)$ is

$$\mathcal{D}(\text{coz}(DIAG(f)), \text{coz}(KRON(g,h))$$

Now we will see that the following identity holds true.

$$TCONV(f,g,h) = CONV(DIAG(f), KRON(g,h))$$

To see this, notice that $TCONV$ can be written as

$$TCONV(f,g,h)(n,m) = \sum_{k=-\infty}^{\infty} F(k,k)g(n-k)h(m-k)$$

where

$$F(n,m) = \begin{cases} f(n) & n = m \\ 0 & \text{otherwise} \end{cases}$$

Thus, we have

$$F = DIAG(f)$$

Next write

$$TCONV(f,g,h)(n,m) = \sum_{k=-\infty}^{\infty} F(k,k)H(n-k, m-k)$$

where

$$H(n-k, m-k) = g(n-k)h(m-k)$$

Comparing this expansion with the definition of the Kronecker product shows that

$$H = g \boxed{K} h$$

It therefore follows that

$$TCONV(f,g,h) = DIAG(f) * KRON(g,h)$$

Because of this relation we can perform the triple convolution using the parallel two dimensional convolution algorithm. Indeed, the entire procedure for performing $TCONV$ is illustrated in Figure 7.4. Additionally, since the usual convolution algorithm is commutative, it follows that

$$TCONV(f,g,h) = KRON(g,h) * DIAG(f)$$

This identity is illustrated in Figure 7.5 and may be simpler to implement then the previous identity.

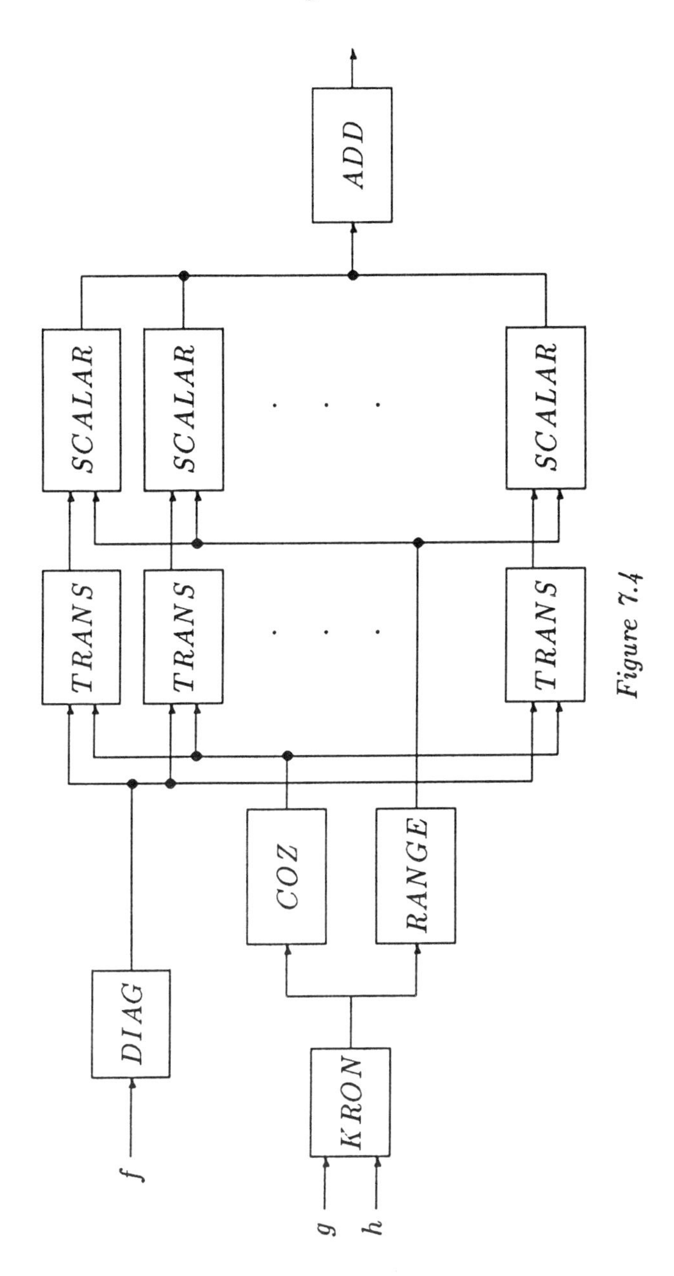
ADD
SCALAR
SCALAR
SCALAR
TRANS
TRANS
TRANS
COZ
RANGE
DIAG
KRON
f
g
h

Figure 7.4

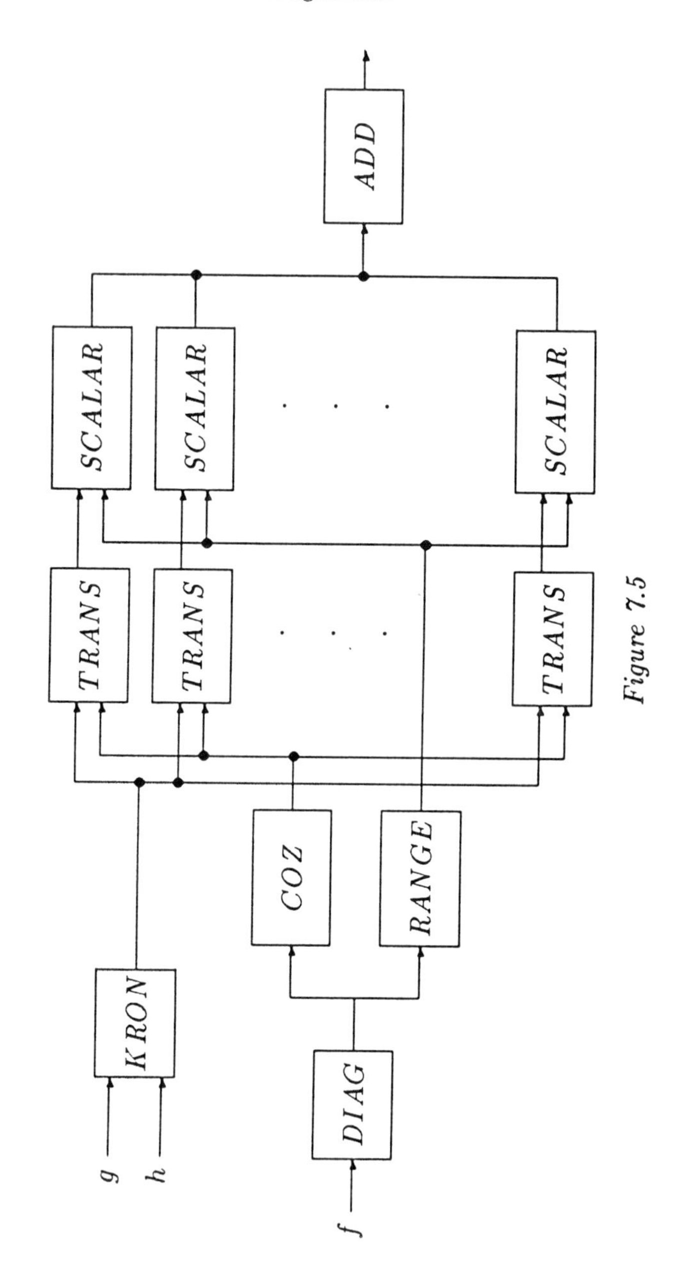

ADD
SCALAR
SCALAR
SCALAR
TRANS
TRANS
TRANS
KRON
COZ
RANGE
DIAG
g
h
f

Figure 7.5

7.6 Volterra Convolution and Volterra Series

Volterra type convolution is an operation involving a one dimensional signal f as input and a one dimensional signal g as output. The output is defined whenever the series below converges:

$$g(n) = \sum_{i_1,i_2,\ldots i_k=-\infty}^{\infty} h^k(i_1, i_2, \ldots, i_k) f(n-i_1) f(n-i_2) \cdots f(n-i_k)$$

Here h^k is a k dimensional real valued structuring signal also defined as the *kernel.* In a single dimension

$$g(n) = \sum_{k=-\infty}^{\infty} h^1(k) f(n-k)$$

and we have the one dimensional convolution. Similarly, in two dimensions we have

$$g(n) = \sum_{k,j=-\infty}^{\infty} h^2(k,j) f(n-k) f(n-j)$$

In three dimensions, the Volterra convolution is given by

$$g(n) = \sum_{k,j,i=-\infty}^{\infty} h^3(k,j,i) f(n-k) f(n-j) f(n-i)$$

We will denote the Volterra convolution of f, using a k dimensional kernel, by $VOLTK(f)$. This is symbolized by the block diagram:

$$f \longrightarrow \boxed{VOLTK} \longrightarrow VOLTK(f)$$

As in previous sections, co-zero sets can be utilized in obtaining more efficient calculations. This will be done for signals of finite support. Indeed, in two dimensions, it is true that

$$g(n) = \begin{cases} \displaystyle\sum_{\substack{(k,j)\in \mathrm{coz}(h) \\ n-k\in \mathrm{coz}(f) \\ n-j \in \mathrm{coz}(f)}} h^2(k,j) f(n-k) f(n-j) & n \in C \\ 0 & otherwise \end{cases}$$

where $\mathcal{C} = \{n \mid (n,n) \in \mathcal{D}(A,B)\}$. Here, as usual, $\mathcal{D}$ is the dilation operation and

$$A = \text{coz}(h), \quad B = \text{coz}(KRON(f,f))$$

Example 7.7 *Suppose that*

$$f = \begin{pmatrix} 2 & -1 \end{pmatrix}_0^0 \quad \textit{and} \quad h^2 = h = \begin{pmatrix} 1 & -1 \\ 2 & 3 \end{pmatrix}_{0,1}^{0}$$

We will find the Volterra convolution. Note that

$$\text{coz}(h) = \{(0,1),(0,0),(1,0),(1,1)\}$$

and $\text{coz}(f) = \{0,1\}$. *Now*

$$f \boxed{K} f = \begin{pmatrix} -2 & 1 \\ 4 & -2 \end{pmatrix}_{0,1}^{0}$$

and setting

$$A = \text{coz}(KRON(f,f)) = \{(0,1),(0,0),(1,0),(1,1)\}$$

$$B = \text{coz}(h) = \{(0,1),(0,0),(1,0),(1,1)\}$$

We then have

$$\mathcal{D}(A,B) = \{(0,1),(0,0),(1,0),(1,1),(0,2),(1,2),(2,2),(2,1),(2,0)\}$$

Accordingly, a support region for g *is* $\{0,1,2\}$. *So*

$$g(0) = h(0,0)f(0)f(0) = 8$$
$$g(1) = h(0,0)f(1)f(1) + h(1,1)f(0)f(0) + h(0,1)f(1)f(0) = -10$$
$$g(2) = h(1,1)f(1)f(1) = -1$$

and

$$g = \begin{pmatrix} 8 & -10 & -1 \end{pmatrix}_0^0$$

The Volterra convolution in dimensions higher than two can be found in an analogous manner. Of course, concepts such as co-zero sets, dilation operation, and Kronecker products are easily extended to the n dimensional case. This is done in a straightforward manner; an example will illustrate the procedure.

Example 7.8 *Let*

$$f = \begin{pmatrix} 2 & -1 \end{pmatrix}_0^0$$

and

$$h^3(k,j,i) = \begin{cases} 1 & (0,0,0) \\ 2 & (1,0,0) \\ -1 & (1,1,1) \\ 0 & otherwise \end{cases}$$

We will find the Volterra convolution. Note that

$$\text{coz}(h) = \{(0,0,0),(1,0,0),(1,1,1)\}$$

and $\text{coz}(f) = \{0,1\}$. *Also note that we use* h *for simplicity in place of* h^3. *A support region for* g *consists of the projections of points along the diagonal of the dilated set obtained from the co-zero set for* h *and the co-zero set of the Kronecker product of the input signal* f. *It is* $\{0,1,2\}$. *So*

$$\begin{aligned} g(0) &= h(0,0,0)f^3(0) = 8 \\ g(1) &= h(0,0,0)f^3(1) + h(1,0,0)f(0)f^2(1) + h(1,1,1)f^3(0) = -5 \\ g(2) &= h(1,1,1) = f^3(1) = 1 \end{aligned}$$

Accordingly, $g = \begin{pmatrix} 8 & -5 & 1 \end{pmatrix}_0^0$.

The Volterra series with input signal f and output signal g is

$$g(n) = \sum_{k=1}^{\infty} \sum_{i_1,i_2,\ldots,i_k=-\infty}^{\infty} h^k(i_1,i_2,\ldots,i_k) \prod_{j=1}^{k} f(n-i_j)$$

Noting that this is an infinite sum of Volterra convolutions implies that

$$g(n) = \sum_{k=1}^{\infty} VOLTK(f)(n)$$

Example 7.9 *Referring to the previous examples where*

$$f = \begin{pmatrix} 2 & -1 \end{pmatrix}_0^0$$

and the Volterra series g *of* f *is to be computed where we assume that*

$$h^1 = 0_Z, \quad h^2 = \begin{pmatrix} 1 & -1 \\ 2 & 3 \end{pmatrix}_{0,1}^{0}$$

$$h^3(k,j,i) = \begin{cases} 1 & (0,0,0) \\ 2 & (1,0,0) \\ -1 & (1,1,1) \\ 0 & otherwise \end{cases}$$

and

$$h^m = 0_{Z \times Z \times \cdots \times Z} \quad for\ m \geq 4$$

Then it is true that g can be found from the block diagram

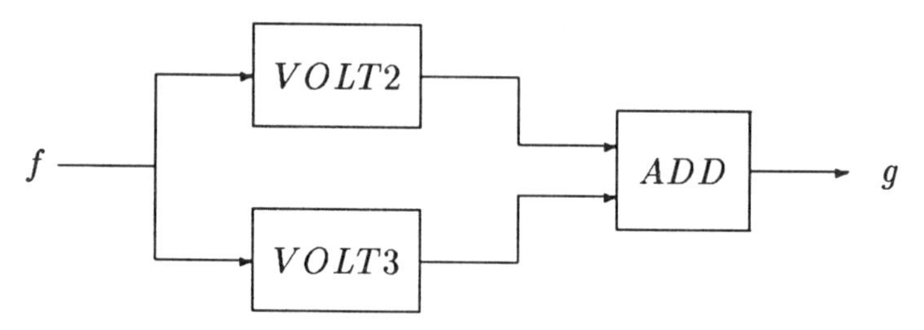

Since

$$VOLT2(f) = \begin{pmatrix} 8 & -10 & -1 \end{pmatrix}_0^0$$

and

$$VOLT3(f) = \begin{pmatrix} 8 & -5 & 1 \end{pmatrix}_0^0$$

then

$$g = \begin{pmatrix} 16 & -15 \end{pmatrix}_0^0$$

A sufficient condition for the convergence of the Volterra series is if each structuring signal h^k is in l_1, and the input signal f is bounded in absolute value by a constant less than one, and if

$$a_k = \sum_{i_1, i_2, \ldots, i_k = -\infty}^{\infty} \left| h^k(i_1, i_2, \ldots, i_k) \right| < \infty$$

This follows since for some real number a, $|a| < 1$,

$$|f(m)| \leq |a| \quad \text{for all } m$$

and

$$|VOLTK(f)| \leq |a_k|\,|a|^k$$

and so

$$|g(n)| \leq \sum |a_k|\,|a|^k \leq \frac{M}{(1 - |a|)}$$

where since

$$a_k \to 0, \quad |a_k| \leq M \quad \text{for all } k$$

Needless to say, a Volterra series consisting of only a finite number of terms involving signals of finite support, always converges.

7.7 Parallel Algorithms for Volterra Convolution

Volterra convolution can be found using the parallel algorithm illustrated in Figure 7.7. This diagram holds true in two dimensions as well as three, four, and so on. We will provide a more in-depth description in the two dimensional case. Since,

$$VOLT2(f)(n) = \sum_{k,j=-\infty}^{\infty} h^2(k,j)f(n-k)f(n-j)$$

we see that if we let

$$F(r,s) = f(r) \boxed{K} f(s)$$

then we can write

$$VOLT2(f)(n) = \sum_{k,j=-\infty}^{\infty} h^2(k,j)F(n-r,n-s)$$

Moreover, if we let

$$q(n,m) = \sum_{k,j=-\infty}^{\infty} h^2(k,j)F(n-r,m-s)$$

then q can be found using the two dimensional parallel algorithm given in Figure 7.7. Additionally,

$$VOLT2(f)(n) = q(n,n)$$

That is,

$$VOLT2(f) = IDIAG(q)$$

This validates the block diagram given in Figure 7.6. A change of variables also shows that the Volterra convolution is commutative and so the parallel algorithm given in Figure 7.7 can also be employed.

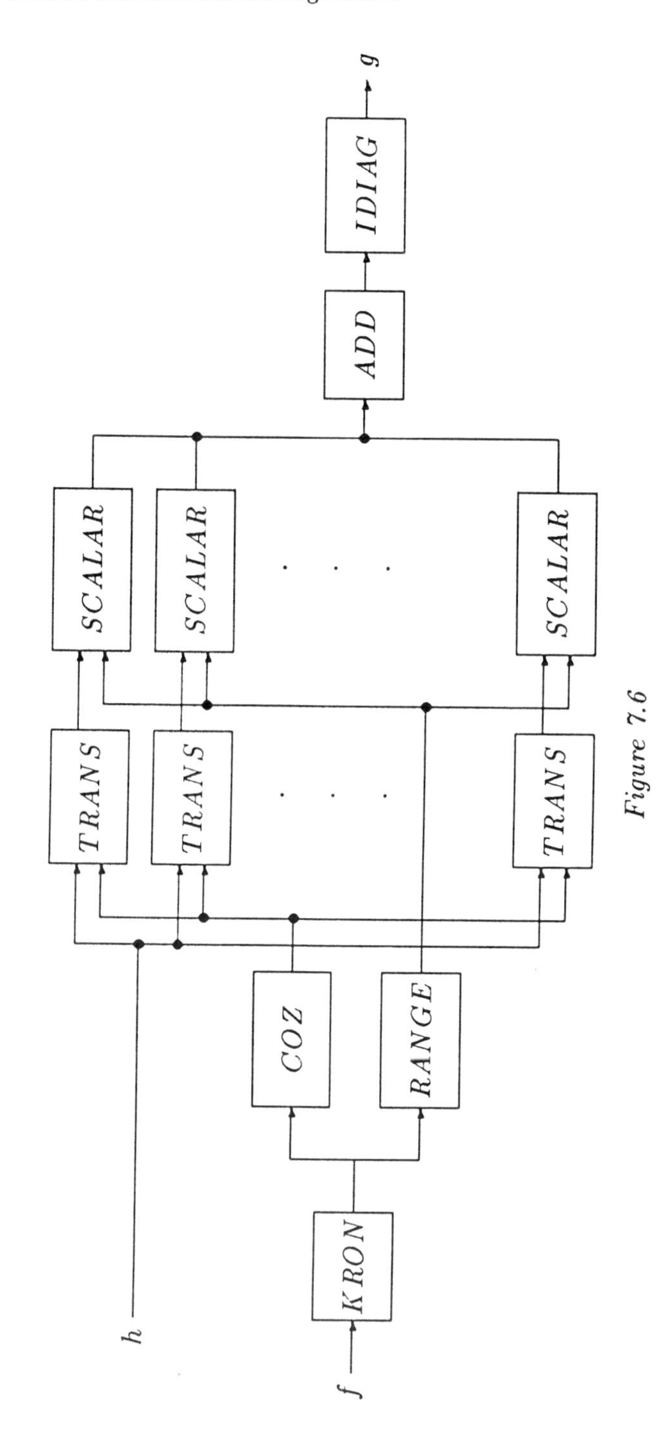
f
h
KRON
COZ
RANGE
TRANS
TRANS
TRANS
SCALAR
SCALAR
SCALAR
ADD
IDIAG
g

Figure 7.6

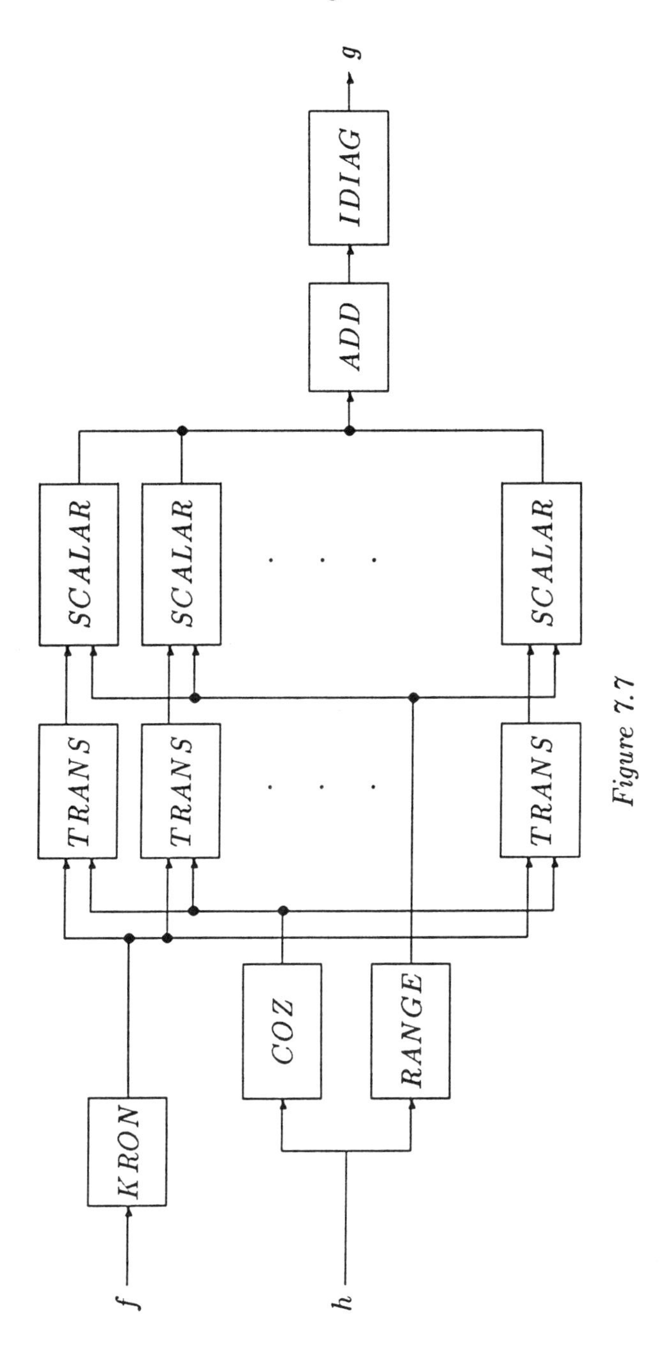
f
h
KRON
COZ
RANGE
TRANS
TRANS
TRANS
SCALAR
SCALAR
SCALAR
ADD
IDIAG
g

Figure 7.7

An example will make the algorithm clearer.

Example 7.10 *We will find the Volterra convolution*

$$g(n) = \sum_{k,j=-\infty}^{\infty} h^2(k,j)f(n-k)f(n-j)$$

using

$$h^2 = h = \begin{pmatrix} 1 & -1 \\ 2 & 3 \end{pmatrix}_{0,1}^{0}$$

and

$$f = \begin{pmatrix} 2 & -1 \end{pmatrix}_{0}^{0}$$

We shall now use the two dimensional parallel algorithm given in Figure 7.6. Forming

$$f \boxed{K} f$$

inputting the resulting signal into the co-zero and range blocks provides parameters by which h is translated and scalar multiplied. This gives

$$-2h_{0,1} = \begin{pmatrix} -2 & 2 & 0 \\ -4 & -6 & 0 \\ 0 & 0 & 0 \end{pmatrix}_{0,0}^{0}$$

$$h_{1,1} = \begin{pmatrix} 0 & 1 & -1 \\ 0 & 2 & 3 \\ 0 & 0 & 0 \end{pmatrix}_{0,0}^{0}$$

$$2h_{0,0} = \begin{pmatrix} 0 & 0 & 0 \\ 4 & -4 & 0 \\ 8 & 12 & 0 \end{pmatrix}_{0,0}^{0}$$

$$-2h_{1,0} = \begin{pmatrix} 0 & 0 & 0 \\ 0 & -2 & 2 \\ 0 & -4 & -6 \end{pmatrix}_{0,0}^{0}$$

These signals are added to give

$$t = \begin{pmatrix} -2 & 3 & -1 \\ 0 & -10 & 5 \\ 8 & 8 & -6 \end{pmatrix}_{0,0}^{0}$$

and finally, the inverse diagonal transform of t is obtained to give the desired result:

$$g = IDIAG(t) = \begin{pmatrix} 8 & -10 & -1 \end{pmatrix}_0^0$$

As mentioned earlier, the parallel algorithm can also be employed in higher dimensions. The next example will make this more convincing.

Example 7.11 *We will find the Volterra convolution*

$$g(n) = \sum_{k,j,i=-\infty}^{\infty} h(k,j,i)f(n-k)f(n-j)f(n-i)$$

using

$$f = \begin{pmatrix} 2 & -1 \end{pmatrix}_0^0$$

and

$$h(k,j,i) = \begin{cases} 1 & (0,0,0) \\ 2 & (1,0,0) \\ -1 & (1,1,1) \\ 0 & otherwise \end{cases}$$

We will employ the parallel algorithm illustrated in Figure 7.7. In order to do this we must form the Kronecker product $f \boxed{K} f \boxed{K} f$.

$$KRON(f,f,f)(k,j,i) = \begin{cases} 8 & (0,0,0) \\ -4 & (0,0,1) \\ -4 & (0,1,0) \\ -4 & (1,0,0) \\ 2 & (1,1,0) \\ 2 & (1,0,1) \\ 2 & (0,1,1) \\ -1 & (1,1,1) \\ 0 & otherwise \end{cases}$$

Accordingly, three dimensional bound matrices can be employed in denoting three dimensional functions. We will leave it to the reader to construct such structures. As in the two dimensional case, the signal h is input into the co-zero and range blocks to obtain:

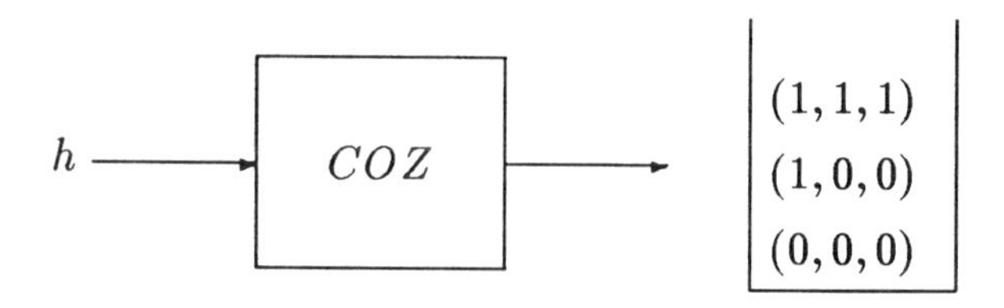

and

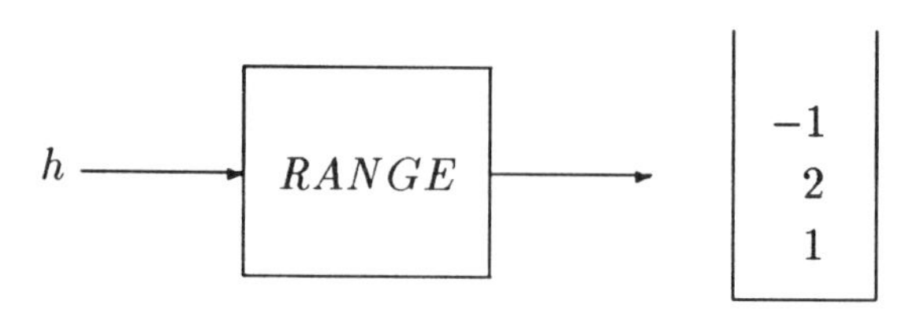

These values are used respectively in translating and scalar multiplying $KRON(f,f,f)$. We will let t denote $KRON(f,f,f)$. The output of the translation scalar connections are

$$1t_{0,0,0} = t$$

$$2t_{1,0,0}(k,j,i) = \begin{cases} 16 & (1,0,0) \\ -8 & (1,0,1) \\ -8 & (1,1,0) \\ -8 & (2,0,0) \\ 4 & (2,1,0) \\ 4 & (2,0,1) \\ 4 & (1,1,1) \\ -2 & (2,1,1) \\ 0 & otherwise \end{cases}$$

and

$$-1t_{1,1,1}(k,j,i) = \begin{cases} -8 & (1,1,1) \\ 4 & (1,1,2) \\ 4 & (1,2,1) \\ 4 & (2,1,1) \\ -2 & (2,2,1) \\ -2 & (2,1,2) \\ -2 & (1,2,2) \\ 1 & (2,2,2) \\ 0 & otherwise \end{cases}$$

Letting

$$s = t + 2t_{1,0,0} - t_{1,1,1}$$

then

$$g = IDIAG(s) = \begin{pmatrix} 8 & -5 & 1 \end{pmatrix}_0^0$$

7.8 Exercises

1. Convolve each of the signals: g^0, g^1, and g^2 where

$$g^0 = \begin{pmatrix} -1 & 1 & 2 \end{pmatrix}_2^0$$

$$g^1 = \begin{pmatrix} -1 & 0 & 2 & 1 \end{pmatrix}_1^0$$

and

$$g^2 = \begin{pmatrix} 1 & 1 & 2 & 3 \end{pmatrix}_2^0$$

with

$$f = \begin{pmatrix} 2 & -1 & 1 \end{pmatrix}_1^0$$

Do this by making g^0, g^1, and g^2 into a single two dimensional signal g, and then use the parallel two dimensional algorithm.

2. Show how the two dimensional parallel correlation algorithm can be employed to correlate g^0, g^1, and g^2 with f.

3. Is it true that

$$DIAG(IDIAG(g)) = g \ ?$$

4. Give an example of two digital signals f and g where $f \neq g$ and for which it is true that

$$f \boxed{K} g = g \boxed{K} f$$

5. Suppose that

$$f = \begin{pmatrix} 2 & 1 \end{pmatrix}_0^0$$

$$g = \begin{pmatrix} 1 & 1 & 2 \end{pmatrix}_2^0$$

and

$$h = \begin{pmatrix} 1 & -1 & 2 \end{pmatrix}_2^0$$

are given. Find the triple convolution for f, g, and h. Do this using the pointwise definition.

6. Perform the triple convolution for f, g, and h, as given in Exercise 5, using the parallel algorithm.

7. Find the triple convolution using a parallel algorithm if

$$f = \begin{pmatrix} 2 & 1 & -1 \end{pmatrix}_0^0$$

and

$$g = h = \begin{cases} 1 & \text{at the even integer points} \\ -1 & \text{at the odd integer points} \end{cases}$$

8. Find the Volterra convolution $VOLT2(f)$ where

$$f = \begin{pmatrix} 2 & -1 & 1 \end{pmatrix}_0^0$$

and

$$h^2 = \begin{pmatrix} 1 & -1 \\ 2 & -2 \end{pmatrix}_{0,0}^0$$

Do this using the pointwise formula.

9. Find the Volterra convolution $VOLT3(f)$ using the pointwise formulation where

$$f = \begin{pmatrix} 2 & -1 & 2 \end{pmatrix}_0^0$$

and

$$h^3(i,j,k) = \begin{cases} 1 & (0,0,0) \\ -1 & (0,1,0) \\ 2 & (1,0,0) \\ 1 & (1,1,1) \\ 0 & otherwise \end{cases}$$

10. Perform $VOLT2(f)$ given in Exercise 8 using the parallel algorithm.

11. Repeat the operation $VOLT3(f)$ given in Exercise 9 using the parallel algorithm.

12. Find the Volterra series g of

$$f = \begin{pmatrix} 2 & -1 & 1 \end{pmatrix}_0^0$$

with kernels

$$h^1 = \begin{pmatrix} 1 & 1 & 1 \end{pmatrix}_3^0$$

$$h^2 = \begin{pmatrix} 1 & -1 \\ 2 & -2 \end{pmatrix}_{0,0}^0$$

h^3 given in Exercise 9 and

$$h^n = 0$$

for $n > 3$.

Appendix

Set Operations and Morphology for Two Dimensional Digital Signal Processing

Set theoretic morphology consists of a mixture of some simple set theoretic operations and arithmetic type operations. The set operations are those of union and intersection of two sets along with subsets of a given set. The arithmetic type operations involve only addition and multiplication. Each of these operations will be described using sets consisting of lattice points. A lattice point is a pair of integers.

We will be interested in the set A, which contains only a finite number of lattice points. The set of all lattice points is denoted by $\mathbf{Z} \times \mathbf{Z}$. The number of elements in a set A in this case is called the cardinality of A and is denoted by card(A). For instance, the set A where

$$A = \{(0,0),(0,3),(4,0),(7,7)\}$$

has card$(A) = 4$. We use the symbol $\in$ to represent membership in a set and the symbol $\notin$ to represent its negation. Thus $(0,3) \in A$ while the point $(1,1) \notin A$.

Whenever every element of a set B also belongs to another set A we say that B is a subset of A. Intuitively, B is a piece of A. This relation is denoted by

$$B \subset A$$

Thus, whenever $x \in B$ implies that $x \in A$ it follows that $B \subset A$. When $x \in B$ does not imply that $x \in A$, then B is not a subset of A. Moreover, two sets are said to be equal when each is a subset of the other. Throughout the Appendix, we are dealing with subsets A of $\mathbf{Z} \times \mathbf{Z}$, thus $A \subset \mathbf{Z} \times \mathbf{Z}$. In general, it is always true that $A \subset A$ and $\emptyset \subset A$, where $\emptyset$ denotes the empty set. It is the set consisting of no lattice points and is also denoted by $\{\}$.

The operation of union of two sets involves combining the elements of the sets. More formally, the union of sets A and B is denoted by $A \cup B$ and is defined by

$$A \cup B = \{x \mid x \in A \text{ or } x \in B\}$$

An arithmetic type operation, called translation involves the addition of a set with a lattice point. Translation is denoted by $A+(n,m)$ where $(n,m) \in \mathbf{Z} \times \mathbf{Z}$ and it is defined by

$$A + (n,m) = \{(p+n, q+m) \text{ with } (p,q) \in A\}$$

The resulting set $A+(n,m)$ is also called the shifting of A by (n,m). This operation entails moving the set A.

The final operation, called the 180° rotation, is the multiplication of a set A by -1, denoted by $-A$ and is defined by

$$-A = \{(-p,-q) \text{ for } (p,q) \in A\}$$

The resulting set is a rotation of A 180° "about the origin."

In set morphology there are two basic operations. The first is called *dilation*, or sometimes called Minkowski addition. Given two finite subsets of integers A and B, the dilation of A by B is denoted by $\mathcal{D}(A,B)$. It is defined by translating A by each element of B and forming the union of all the resulting translates.

$$\mathcal{D}(A,B) = \bigcup_{(n,m)\in B} (A + (n,m))$$

If we let

$$A = \{(0,0),(0,2),(3,0),(5,5)\}$$

and

$$B = \{(-1,1),(2,1)\}$$

To find the dilation of A by B we first find all translates of A by elements in B:

$$A + (-1,1) = \{(-1,1),(-1,3),(2,1),(4,6)\}$$

$$A + (2,1) = \{(2,1),(2,3),(5,1),(7,6)\}$$

Next, form the union which provides the desired result

$$\mathcal{D}(A,B) = \{(-1,1),(-1,3),(2,1),(4,6),(2,3),(5,1),(7,6)\}$$

In general, the order in which the union is performed is irrelevant. This follows since the union operation satisfies the commutative and associative laws, i.e. we respectively have

$$A \cup B = B \cup A$$

$$(A \cup B) \cup C = A \cup (B \cup C)$$

As a consequence, the dilation operation also satisfies the commutative and associative laws, i.e. respectively we have

$$\mathcal{D}(A,B) = \mathcal{D}(B,A)$$

$$\mathcal{D}\left(\mathcal{D}(A,B),C\right) = \mathcal{D}\left(A,\mathcal{D}(B,C)\right)$$

The second basic operation in morphology is *erosion*, it too is a binary operation. Thus, given two subsets of lattice points A and B, the erosion of A by B will produce a subset of lattice points. The resulting set, denoted by $\mathcal{E}(A,B)$ is found by first determining all translates of B which are a subset of A. Once this is done, the second and final step is to form the set consisting of the translation values. It is $\mathcal{E}(A,B)$. Thus,

$$\mathcal{E}(A,B) = \{(n,m) \mid B + (n,m) \subset A\}$$

When performing the erosion of A by B, the second argument is called the structuring element. The second argument when performing dilation is also called the structuring element. For a given structuring element, these operations are somewhat opposites. While the dilation of a set A by a structuring element B expands the set A in some sense, the erosion operation performed on A, using the structuring element B, shrinks the set A. As an example, suppose that

$$A = \{(0,2),(1,2),(2,2),(5,3),(6,3),(7,3),(8,4)\}$$

and

$$B = \{(-1,1),(0,1)\}$$

In order to find $\mathcal{E}(A,B)$ we first determine all translates of B which are a subset of A. So

$$B + (1,1) = \{(0,2),(1,2)\} \subset A$$

$$B + (2,1) = \{(1,2),(2,2)\} \subset A$$

$$B + (6,2) = \{(5,3),(6,3)\} \subset A$$

$$B + (1,1) = \{(6,3),(7,3)\} \subset A$$

are the only translates of B which are subsets of A. Next we form a set consisting of the translation values $(1,1)$, $(6,2)$, and $(7,2)$ applied to B. Thus,

$$\mathcal{E}(A,B) = \{(1,1),(2,1),(6,2),(7,2)\}$$

Unlike the dilation operation, erosion in general does not satisfy the commutative or associative laws. The following identity holds true:

$$\mathcal{E}(\mathcal{E}(A,B),C) = \mathcal{E}(A,\mathcal{D}(B,C))$$

The erosion operation can also be found using Minkowski subtraction. The operation is denoted by $A \ominus B$ and is defined as

$$A \ominus B = \bigcap_{(n,m)\in B} A + (n,m)$$

Thus, $A \ominus B$ is similar to $A \oplus B$, except that an intersection is employed in place of the union operation. We also have

$$\mathcal{E}(A,B) = A \ominus -B$$

The third morphological operation we shall encounter is the opening operation. The opening of A by B, denoted by $\mathcal{O}(A,B)$ is found by forming the union of all translates of B which are subsets of A. Thus,

$$\mathcal{O}(A,B) = \bigcup_{B+(n,m)\subset A} (B + (n,m))$$

As in the discussion of erosion, B is said to be the structuring element when forming $\mathcal{O}(A,B)$. Also, as in the erosion operation, the result cannot increase in cardinality when it is opened by the structure element B. For example, suppose that

$$A = \{(-1,0),(2,0),(4,1),(5,1),(7,2),(8,2)\}$$

$$B = \{(-1,2),(0,2)\}$$

and $\mathcal{O}(A,B)$ is to be found. First, we find all translates of B which are subsets of A.

$$B + (5,-1) = \{(4,1),(5,1)\} \subset A$$

$$B + (8,0) = \{(7,2),(8,2)\} \subset A$$

This step is similar to the first step in finding $\mathcal{E}(A,B)$, however the final step is different. We must union those translates of B which are subsets of A. Thus,

$$\mathcal{O}(A,B) = \{(4,1),(5,1),(7,2),(8,2)\}$$

The opening operation satisfies various equational constraints. Among them we mention that the opening is antiextensive, increasing, and idempotent. Respectively the following holds true:

$$\mathcal{O}(A,B) \subset A$$

$$A \subset D \Rightarrow \mathcal{O}(A,B) \subset \mathcal{O}(D,B)$$

and

$$\mathcal{O}(\mathcal{O}(A,B),B) = \mathcal{O}(A,B)$$

The first expression above shows that the opening operation produces a subset of A. The second property intuitively says that the opening of a larger set is larger than the opening of a smaller set within the original set. The final relation says that successive openings by the same structuring element results in the same set.

The final morphological operation is the closing of A by B. This operation is denoted by $\mathcal{C}(A,B)$, and again B is often called the structuring element. The lattice point (n,m) is in $\mathcal{C}(A,B)$ if and only if

$$(B + (p,q)) \cap A \neq \emptyset$$

for any $B + (p,q)$ containing (n,m). Thus, in order to find $\mathcal{C}(A,B)$ we find only those lattice points which have the property that every translate of B containing them also has nonempty intersection with A. It follows that if $(n,m) \in A$ then $(n,m) \in \mathcal{C}(A,B)$. That is the closing operation is extensive, i.e. $A \subset \mathcal{C}(A,B)$. Additionally, like the opening operation, the closing operation is increasing and idempotent. Respectively it is true that if

$$A \subset D \quad \text{then} \quad \mathcal{C}(A,B) \subset \mathcal{C}(D,B)$$

and

$$\mathcal{C}(\mathcal{C}(A,B),B) = \mathcal{C}(A,B)$$

Looking at a simple example, suppose that

$$A = \{(0,2),(4,2),(6,2)\}$$

and

$$B = \{(0,2),(1,2)\}$$

To find the closing of A by B we must find all lattice points (n,m) having the property that every translate of B containing (n,m) also contains a point from the set A. We know that

$$(0,2) \in \mathcal{C}(A,B)$$

$$(4,2) \in \mathcal{C}(A,B)$$

and

$$(6,2) \in \mathcal{C}(A,B)$$

since these points are in A. Now every lattice point is located in just two translates of B since B has cardinality two. The point $(n,m) = (5,2)$ is located in

$$B + (4,0) = \{(4,2),(5,2)\}$$

and $(n,m) = (5,2)$ is located in

$$B + (4,0) = \{(5,2),(6,2)\}$$

Moreover,

$$(B + (4,0)) \cap A = \{(4,2)\} \neq \emptyset$$

$$(B + (5,0)) \cap A = \{(6,2)\} \neq \emptyset$$

and so $(5,2) \in \mathcal{C}(A,B)$. Thus, we have

$$\mathcal{C}(A,B) = \{(0,2),(4,2),(5,2),(6,2)\}$$

Both the opening and closing operations can be obtained from the basic operations of dilation and erosion. Specifically, we have

$$\mathcal{O}(A,B) = \mathcal{D}(\mathcal{E}(A,B),B)$$

and

$$\mathcal{C}(A,B) = \mathcal{E}(\mathcal{D}(A,-B),-B)$$

References

[1] Giardina, C. R., "A Many Sorted Algebra Based Signal Processing Data Flow Language," *Fifth International Conference on Mathematical Modeling*, September 1985.

[2] Banks, S. P., *Signal Processing, Image Processing and Pattern Recognition*, Prentice Hall, NJ, 1990.

[3] Bochner, S. and Martin, W. T., *Several Complex Variables*, Fifth Printing, Princeton University Press, NJ, 1967.

[4] Castleman, K. R., *Digital Image Processing*, Prentice Hall, NJ, 1979. International, UK, 1991.

[5] Dougherty, E. R. and Giardina, C. R., *Matrix Structured Image Processing*, Prentice Hall, NJ, 1987.

[6] Dougherty, E. R. and Giardina, C. R., "Bound Matrix Structured Image Processing," *1986 Conference on Intelligent Systems and Machines,* Published in Conference Proceedings, Oakland Univ., Rochester, Michigan.

[7] Giardina, C.R. and Dougherty, E. R., *Morphological Methods in Image and Signal Processing*, Prentice Hall, NJ, 1988.

[8] Grove, A.C., *An Introduction to the Laplace and The $\mathcal{Z}$-Transform*, Prentice Hall

[9] Hochschild, G., *The Structure of Lie Groups*, Holden-Day, San Francisco, 1965.

[10] Jacobson, N., *Lie Algebras* , Dover Publications, NY, 1979.

[11] Jury, E.I., *Theory and Application of The $\mathcal{Z}$-Transform Method,* Krieger Publishing Co., FL, 1986.

[12] Kolmogorov, A. N. and Fomin, S.V., *Elements of The Theory of Functions and Functional Analysis*, Vol. 1, Metric and Normed Spaces, Graylock Press, Rochester, NY, 1957.

[13] Lim, Jae S., *Two-Dimensional Signal and Image Processing*, PTR Prentice Hall, NJ, 1990.

[14] Rudin, W., *Functional Analysis*, McGraw-Hill, NY, 1973.

Index

A

ABS operation, 24
ADD operation, 17
annulus, 129

B

bound matrix, 8

C

Cauchy-Buniakovski-Schwarz inequality, 104
CLIP operation, 28
co-zero set, 59
convolution, 60
convolution theorem
  finite support, 126
  summable functions, 144
  Banach algebra properties, 79
  bound matrix for, 65
  finite support, 61
  non-finite support, 73
  parallel algorithm, 68
correlation, 103
correlation theorem, 146
cyclic group, 198

D

DIAG transformation, 252
difference equation
  homogeneous, 169
  non-homogeneous, 169
  Z transform method, 179
difference equations, 159
DIFLIP (diagonal reflection) operation, 48
digital signal, 1
dilation, 61
discrete Fourier transform (DFT), 230
domain induced operations, 35

F

filtering, 83

H

high-pass filter, 89
HOR (horizontal) operation, 48

I

inverse discrete Fourier transform (IDFT), 233
inverse Z transform, 117

K

Kronecker Product (KRON), 254

L

lattice points, 1
Laurent series, 129
low-pass filter, 84

M

macro, 20
MAX operation, 19
MIN operation, 23
minimal bound matrix, 12
Minkowski addition (dilation), 205

MINUS operation, 21
modulo addition, 196
morphological operations
    closing, 209
    dilation, 205
    erosion, 206
    opening, 208
moving average filter, 84
MULT operation, 18

N

NINETY operation, 38

O

OFFSET operation, 26

P

pattern matching, 107

R

range induced operations, 17
REFLECT operation, 40
Reinhart domain, 129

S

SCALAR operation, 26
shift operation, 35, 198
SQ operation, 20
support region, 10

T

terms, 20
TRAN (translation) operation, 45
transfer function, 148
triple convolution (TCONV), 259

V

VERT (vertical) operation, 48
Volterra convolution, 265

W

wraparound convolution, 212
wraparound convolution theorem, 228
    parallel algorithm, 216
wraparound correlation, 218
wraparound correlation theorem, 229
    parallel algorithm, 221
wraparound set morphology, 204
wraparound signals, 187
    co-zero set, 190
    domain induced operations, 195
    range induced operations, 190
    successor function, 195
wraparound Z transforms, 223

Z

Z transform, 115
    finite support, 125